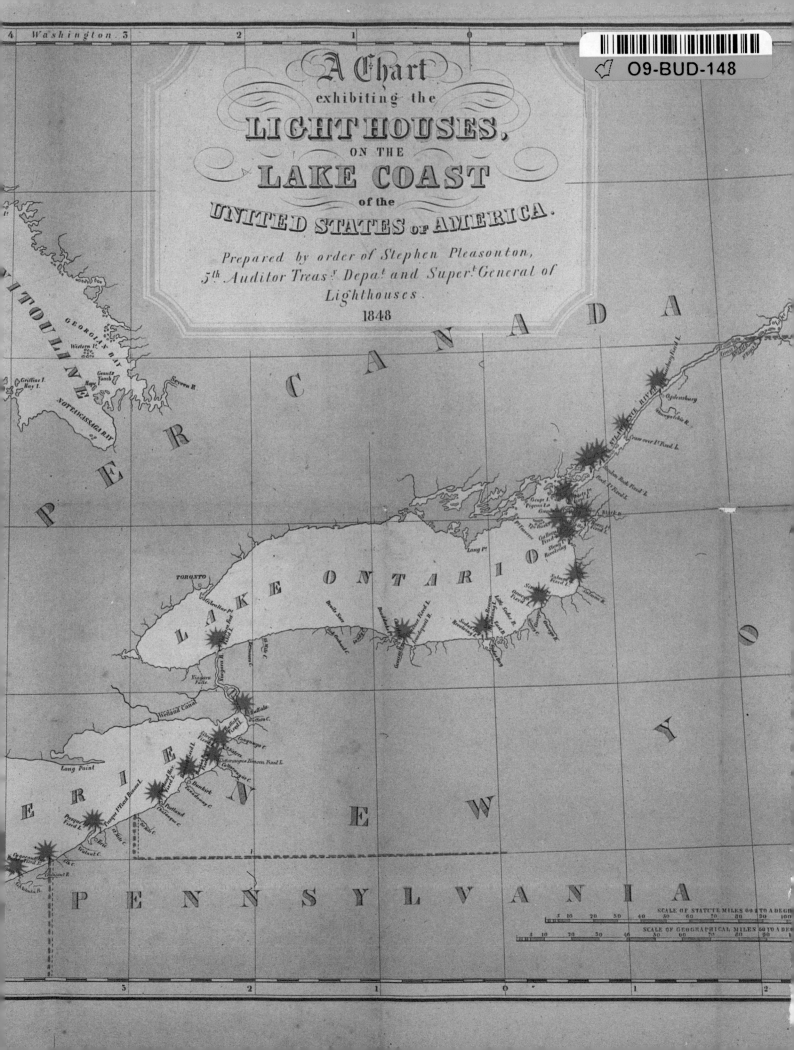

A Chart
exhibiting the
LIGHT HOUSES,
ON THE
LAKE COAST
of the
UNITED STATES OF AMERICA.

Prepared by order of Stephen Pleasonton,
5th Auditor Treasy Depat and Supert General of
Lighthouses.
1848

The Northern Lights

Lighthouses of The Upper Great Lakes

BY CHARLES K. HYDE

COLOR PHOTOGRAPHS BY ANN AND JOHN MAHAN

TwoPeninsula Press • Lansing, Michigan • 1986

EDITOR: Richard Morscheck.
BOOK DESIGN: Joseph E. Sienkiel.
COLOR SEPARATIONS: Lakeland Litho-Plate,
    Inc., Detroit.
PRINTING: Dickinson Press, Grand Rapids.
BINDING: John H. Dekker & Sons, Grand
    Rapids.
PAPER: 70-pound Black and White Gloss by the
    Mead Corporation, Escanaba.
TYPE: Text set in Baskerville.

Library of Congress Catalog Number: 85-062139.
ISBN Number: 0-941912-09-4.

The Northern Lights is Volume VI of the Michigan
Heritage Series, and is a publication of
TwoPeninsula Press, a unit of

Box 30034, Lansing, MI 48909

An early twentieth century photograph of the crew stationed at Point Betsie Light on Lake Michigan. Along with the light at Sherwood Point, Wisconsin, Point Betsie was the last manned lighthouse to be automated in 1983. Photo courtesy U.S. Coast Guard.

EDITOR'S PREFACE

Ever since we first announced our plans to publish this book a little more than two years ago, reader interest has been very strong, and very positive. At the time, we were a little surprised by this enthusiastic response, and it made us wonder what it is about lighthouses that causes people to respond to these man-made structures in the various ways that they do. A short time later, we were also surprised to learn that inquiries about lighthouses were one of the most frequently made requests for information received by the State of Michigan Library. Again we wondered. But as we continued our labors to create the design of this book, to making the final selection of color photographs from more than 5,000 submitted by the photographers, to editing and typesetting the manuscript, to gathering as many historical photos as we could find, to finally putting it all together into the finished product that you now hold in your hands, we began to understand the significance that these aids to navigation have played in the human reach to expand and develop our world. In reading it, we hope you will too.

This book is essentially a narrative of Michigan and Great Lakes maritime history. It is the sixth volume in our Michigan Heritage Series which we began six years ago to celebrate the rich, natural and cultural heritage that makes Michigan the unique and special place that it is and, makes us, the people who live, work, and play here, proud to say ''I'm from Michigan.'' In developing this series we viewed our state history as largely a history of our resources. What was here first was fur-trapping, logging, farming, and mining. Thus, the series is intended to be as much a history of resources as it is a social history. And because these northern lights helped to open this region to settlement, and supported the growth of commercial trade, and, in this century, continued to light the way for thousands of recreational boaters, it is indeed fitting that this book be published to carry on the tradition of excellence established by the earlier volumes. We encourage you to purchase the entire set for your home library. We are confident this book, as well as the other five books in our series, will delight and entertain you and your entire family for many years to come.

Richard Morscheck
May, 1986

CONTENTS

ACKNOWLEDGMENTS

The roots of this work extend back nearly a decade, when I first became interested in lighthouses on the Great Lakes. In 1974-77, I completed an inventory of Michigan's historic engineering buildings and structures, including lighthouses, for the Historic American Engineering Record, National Park Service. I later conducted an historical survey, along with Carol Poh Miller of Cleveland, of United States Coast Guard properties on the Great Lakes. The cooperation of James Woodward of the Great Lakes Historical Society was essential to that work. After a brief hiatus, I then began work on this book, accumulating an entirely new set of debts in the process.

Several individuals have given me invaluable assistance in locating research materials. Charles Wallin allowed me to examine materials collected by his mother, Helen Wallin, who worked for the Michigan Department of Natural Resources for many years and had a lifelong interest in Michigan lighthouses. Dr. Martha Bigelow, Director of the Michigan History Division, Michigan Department of State, gave me access to the Division's extensive historic site files. LeRoy Barnett and John Curry at the State of Michigan Archives, were most helpful in locating photographs and other materials in that repository. David Armour of the Mackinac State Park Commission, Michigan Department of Natural Resources, first brought to my attention a collection of lighthouse drawings now in the State Archives. Carol Dubie of the National Register of Historic Places, National Park Service, provided me with information and encouragement. Finally, Wallace Szumny cheerfully did part of the detailed research on individual lights.

I am also heavily indebted to those who helped improve this book as it developed. Tricia Kessel deciphered and typed several versions of the manuscript. Donn Werling read portions of the book and made useful comments. I am indebted to Mike and Darla Van Hoey, who read the entire manuscript and made many suggestions. Finally I want to thank Russell McKee, former Editor, and Richard Morscheck, Managing Editor, of the *Michigan Natural Resources* Magazine, for their patience and encouragement. None of these individuals, however, are responsible for any errors or omissions in this book. That responsibility is entirely mine.

INTRODUCTION

When my New England relatives expressed surprise that there were lighthouses on the Great Lakes, I could easily forgive them, but I have always found it hard to understand my Michigan friends and colleagues who express the same sentiments. Atlantic and Pacific coast lighthouses are well-known in part because they are often the subject of books and postcards. Great Lakes lighthouses, comparatively speaking, have generally been neglected and ignored. When Russell McKee asked me several years ago to consider writing a book on Great Lakes lighthouses, for the Michigan Heritage Series, I was already primed for the project.

This book gives the reader a general history of the U.S. Lighthouse Service and its descendants, as they functioned on the Great Lakes; a brief summary of the changing design of lighthouses and the equipment that produces their beacons; and finally, a history of the lighthouse keepers and their families. The bulk of the volume outlines the histories of more than one hundred and sixty individual lighthouses. We hope that this book will encourage everyone living in the Great Lakes area to learn more about our lighthouses and other aspects of the region's rich maritime history. There are dozens of excellent museums in the region, with marvelous collections relating to navigation on these "inland seas."

This is not intended to be a comprehensive history of all the lighthouses ever built on the Great Lakes. We do not consider Canadian lighthouses, which merit a separate study. This book looks only at the American lights of the Upper Great Lakes, but does include a few located in the western end of Lake Erie, since these are accessible to Michiganders. We deliberately decided not to write about the hundreds of lights that no longer exist.

This volume is both a history of and a guide to the lighthouses on the Great Lakes that people can still *view,* and, in many cases, can even visit the interiors. The remarkable beauty produced by the man-made structures in their Great Lakes settings can be seen in the outstanding color photographs Ann and John Mahan have created for this book. We hope that it will spur readers to see these lighthouses in person.

Charles K. Hyde
Royal Oak, Michigan

1 FOUNDATIONS AND THE FIRST LIGHTS ON THE GREAT LAKES

Throughout human history, navigation and civilization have gone hand in hand. Starting with the oldest advanced societies of the Middle East, water has served as mankind's first highway and has remained to this day an important avenue for moving goods and people over long distances. Early man used rivers effectively with nothing more advanced than dugout canoes and crude rafts made from bundles of reeds, logs, or inflated animal skins. The earliest use of sails, around 4,000 B.C., was on rafts plying the Nile, Tigris and Euphrates rivers. By about 2,000 B.C., carpenters in Phoenicia and on the Isle of Crete built large sailing ships which opened up the Mediterranean Sea to long-distance trade. The large empires created by the Phoenicians, Cretans, and Greeks in the Eastern Mediterranean were based on long-distance maritime trade and naval power.

Lighthouses and long-distance navigation developed together. The earliest lights were bonfires built on hillsides to guide ships returning at night. Sailing ships reached port at unpredictable hours and their captains naturally preferred to keep sailing as long as the winds held. The Emperor Ptolemy, ruler of Egypt and a successor of Alexander the Great, ordered the erection of what was probably the world's first permanent lighthouse on the Island of Pharos at the entrance to the harbor serving Alexandria. The Hellenistic (Greek) architect Sostratus of Cnidus designed the structure, which went into service in 285 B.C. It stood about four hundred feet tall and used an open fire as the light source. Throughout most of the period of the Roman Empire, Alexandria was the shipping point for huge amounts of grain destined for Rome. The Pharos light survived nearly 1,500 years till an earthquake toppled it. Some historians also believe that the huge statue of Helios (Apollo), better known as the Colossus of Rhodes (270 B.C.), served as a lighthouse, with a fire burning in the statue's hand or eyes. At least one and possibly two of the Seven Wonders of the Ancient World were light-

houses. The Romans built at least thirty substantial lights on the Mediterranean Sea and the Atlantic Ocean before the Empire fell into decline starting in the third century A.D. At Ostia, the great port city serving Rome and located about twenty miles downstream on the Tiber River, they erected a lighthouse building which was more than one hundred feet tall and served as a fortress as well as a beacon.

Lighthouses remained in use during the Middle Ages, but few new ones were built until trade revived around 1100 A.D., when several Italian city-states erected impressive light structures. The uncle of Christopher Columbus served as the keeper at the Genoa lighthouse shortly before his nephew's famous voyage of discovery. During the sixteenth and seventeenth centuries, the Atlantic trade expanded greatly and touched off a wave of new lighthouse construction in France, England and the Low Countries.

At the end of the seventeenth century, the British government awarded Henry Winstanley a contract to design and build a lighthouse on the Eddystone Rocks, a major hazard to navigation located in the English Channel near Plymouth. Construction began in 1696 and Winstanley completed an eighty foot wooden structure two years later. He then raised the tower to a new height of 120 feet the following year. In November, 1703, Winstanley and a repair crew died in the lighthouse when a savage storm demolished the structure. John Rudyerd soon erected a new stone tower sheathed with timber planks in 1708, but it subsequently burned to the water in December, 1755. The civil engineer John Smeaton then designed a new stone tower, which stood from 1759 until 1882, when a new tower replaced it. Robert Stephenson completed a similar stone tower in 1811 on Bell Rock, a dangerous reef in the Firth of Forth, on the eastern coast of Scotland. This tower and the one at Eddystone showed that lighthouses could be built in extremely inhospitable locations. By 1800, the nations of western Europe

View on Lake Huron, Mich. D.

Preceding page, Fort Gratiot Light was the first lighthouse on Lake Huron, built in 1825. But the eighty-six foot tower that stands today was erected in 1861. Above, an unknown artist's nineteeth century watercolor painting of the Fort Gratiot Light. Photo courtesy Dossin Great Lakes Museum. Above right, a 1945 photograph of the Fort Niagara, New York, Light, the first American aid to navigation on the Great Lakes. The original structure, built in 1818, was a wooden tower attached to the mess-house at the fort. Photo courtesy The Great Lakes Historical Society.

maintained more than two hundred major lighthouses. Eddystone also inspired one of the earliest folk songs featuring a lighthouse:

> *My father was the keeper of the Eddystone light,*
> *And he slept with a mermaid one fine night.*
> *From this union there came three,*
> *A porpoise, a porgy, and the other was me.*
> *Yo ho ho, the wind blows free;*
> *Oh, for a life on the rolling sea.*

COLONIAL AMERICAN LIGHTS

The first settlements in the American colonies depended heavily on water-borne trade with England and other European nations. Colonists used bonfires and lanterns to guide ships right from the start, but the first major lighthouse in the colonies and probably in North America dates from 1716. Three years earlier,

Boston area merchants petitioned the Massachusetts colonial legislature, the General Court, to erect a lighthouse and this body appropriated the funds in June, 1715. They built a stone structure on Little Brewster Island, at the Boston Harbor entrance, in about fifteen months and the light was first exhibited on September 14, 1716. The General Court recovered the costs of building and maintaining the lighthouse by charging dues amounting to "one Penney per Ton Inwards, and another Penney Outwards, except Coasters, who are to pay Two Shillings each, at their clearance Out, and all Fishing Vessels, Wood Sloops, etc., Five Shillings each by the Year." George Worthylake served as the first keeper. In 1719, the authorities installed a cannon "to answer Ships in a Fog," the first fog signal placed in service in the New World. The Boston tower suffered damage from an accidental fire in 1751 and from one set by colonial troops in 1775 in an attempt to prevent the British from using it. The attempt failed, and the tower retained its military importance until the redcoats blew it up as they retreated from Boston in June, 1776. Parts of the base survived and served as the foundation for a new tower the Commonwealth of Massachusetts built there in 1783.

The colonists built only ten additional lighthouses prior to 1776, with eight of these located in Delaware and points north. State or local governments bore the construction costs and generally undertook the project only after considerable pressure from local merchants and mariners. Light dues typically supported these colonial lighthouses. The town of Nantucket on the island of the same name built a lighthouse on Brant Point in 1746. Six different structures stood there between 1746 and 1700, three destroyed by storm and two by fire. Rhode Island built the Beavertail Light (1749) on Conanicut Island in Narragansett Bay, while the harbor at New London, Connecticut, had a permanent beacon by 1760. The only colonial lighthouse to survive to the present essentially unchanged is the Sandy

Hook, New Jersey, Light (1764) at the entrance to New York Harbor. The Colonial Assembly of New York financed the light with the profits from four lotteries and the octagonal brick tower withstood British and colonial efforts to destroy it during the American Revolution. The Cape Henlopen Light (1767) at the entrance to Delaware Bay, the water route to Philadelphia, suffered a worse fate. It was built on sand and toppled during a violent storm in 1926.

Two southern lighthouses were latecomers to the list of colonial lights. The Morris Island Light (1767) in Charleston Harbor survived until the Confederates destroyed it in 1861. Georgia's colonial government erected a brick tower on Tybee Island at the entrance to the waterway leading to Savannah in 1771, replacing a tower built there in 1736 to hold a daymark. Finally, three towers in the general vicinity of Boston complete the roster of significant colonial aids to navigation: the Plymouth Light (1769) on Gurnet Point in Plymouth Bay, south of Boston; the Cape Ann Light (1771) on Thatcher's Island at the northern end of Massachusetts Bay; and farther north, the Portsmouth, New Hampshire, Light (1771) built at the harbor entrance.

At the conclusion of the Revolutionary War in 1783, the states repaired existing lighthouses, built some new structures and retained responsibility for their operation. The Commonwealth of Massachusetts completed a new light at Great Point on Nantucket

Island in 1784 and another at Newburyport Harbor near the New Hampshire border in 1788. The states began building three additional lights that were eventually completed by the new Federal government. The unfinished Portland Head Light at Portland, Maine, became Federal property in 1790 and went into service in 1791. The Virginia colony began work on the Cape Henry Light at the entrance to Chesapeake Bay in 1774, but the war stopped all building activity. The State of Virginia then turned the property over to the Federal government in 1789, construction resumed in 1791, and the new light went into service in October 1792. Finally, the State of North Carolina began erecting a light at Bald Head at the mouth of the Cape Fear River, but the Federal government also took over this project and opened the new lighthouse in 1795. By that time, the construction and operation of lighthouses was almost exclusively under Federal government control.

EARLY YEARS OF FEDERAL RESPONSIBILITY, 1789-1820

The new Congress of the United States established Federal authority over lighthouses on August 7, 1789, through the ninth law enacted by the new body and established the principles that aids to navigation would be supported from general revenues and could be used by all vessels free of any charges. Congress gave authority over lighthouses to the Treasury Department,

which retained control until 1903, when jurisdiction went to the Commerce Department. The Secretary of the Treasury directly supervised lighthouse operations from 1789 to 1792 and again from 1802 through 1813, but delegated this responsibility to the Commissioner of Revenue from 1792 to 1802 and from 1813 to 1820.

The states ceded their lighthouses to the Federal government by 1797 and by the turn of the new century, a total of twenty-five were in service. Only four of these were located south of Delaware. The Federal government built nineteen additional lights from 1800 to 1810 and a dozen more from 1811 to 1820, including one in Louisiana and two on the Great Lakes. By 1820, the United States operated 55 lighthouses, three fog signals and 156 buoys, nearly all concentrated on the Atlantic seaboard from Maine to Chesapeake Bay.

Some of the early light towers were wooden, but a more typical design was a round or octagonal tower of brick or stone, standing between 50 and 80 feet high, surmounted by a round or octagonal "lantern," the chamber at the top of the lighthouse which encases and protects the light source. Ancient lighthouses used a wood fire, but coal became a popular fuel by the seventeenth century, in large part because it burned slowly. Unfortunately, coal fires were not very bright and often left a coating of soot on the inside of the glass lantern panels. Many lighthouses, including Eddystone, used candles, but their light was seldom bright enough

to be useful. Significant advances in lighting did not occur until the development of efficient oil-burning lamps in the eighteenth century. The so-called "spider lamp" equipped with four flat wicks came into use in the United States in 1790, but gave off acrid fumes and lots of smoke.

The first lighting breakthrough was the Argand lamp (1781), which featured a hollow circular wick that burned with little or no smoke. An American ship captain, Winslow Lewis, further developed this process and in 1810 patented a lighting system using the Argand lamp combined with a parabolic reflector which intensified and directed the light. In 1812, at the urging of Secretary of the Treasury Albert Gallatin, Congress appropriated $60,000 to buy Lewis's patent rights and award him a contract to convert the nation's lighthouses to his new system. Lewis completed the job in three years. Sperm oil from whales was the illuminant used in American lighthouses through the 1850s. There were a few tests done with alternate fuels, including porpoise oil at the Cape Hatteras Lighthouse in 1803 and 1804 and "earth oil" (kerosene) in 1807. A lengthy experiment with natural gas at the Portland, New York, Light on Lake Erie begun in 1829 proved unsuccessful. The Winslow Lewis system with sperm oil as the illuminant would continue to be the standard means of lighting American lighthouses from 1815 until 1852.

Opposite page, other early lights on the Great Lakes were at Buffalo (1828), top left, and at Cleveland (1829), top right. Both photos courtesy The Great Lakes Historical Society. Above, an early twentieth century photograph of the DeTour Light at the entrance to the St. Mary's River. This is how the station looked after it was rebuilt in 1861. Seventy years later, a new structure was built by the Coast Guard at the end of DeTour Reef about one mile offshore.

THE FIRST GENERATION OF GREAT LAKES LIGHTS, 1818-1852

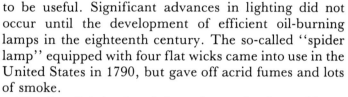

Before the thirteen American colonies gained their independence, the Great Lakes basin was inhabited by Indians and a handful of soldiers at a dozen outposts. The French established Detroit in 1701, but most of the lands bordering the Lakes were not inhabited by whites until the early nineteenth century. As late as 1800, only about 50,000 people lived in the Old Northwest, consisting of the present states of Ohio, Michigan,

Indiana, Illinois, and Wisconsin. The vast majority of these pioneers lived in Ohio and this was still the case in 1820, when the entire region had 800,000 residents and the nation had a total population of slightly less than ten million. The completion of the Erie Canal in 1825, linking Buffalo on Lake Erie with New York City via the Hudson River, marked the start of enormous growth in the Great Lakes region. Settlers could move to the frontiers more readily and the products of the area could be shipped cheaply by water to the cities of the eastern seaboard. Initially, ships could use Lakes Erie, Huron and Michigan with few problems, because they are at virtually the same levels. Lake Erie is about eight feet lower than Lake Huron, but the drop is gradual, running the length of the St. Clair River, Lake St. Clair, and the Detroit River. The large difference in the water levels of Lake Erie and Lake Ontario, some 330 feet, made navigation between the two impossible until the Welland Canal opened in 1829. Through the rest of the nineteenth century, however, the relatively shallow depth (fourteen feet) of this canal severely restricted its value. Similarly, the water surface of Lake Superior is twenty-one feet above that of Lake Huron. The St. Mary's Falls Ship Canal (the Soo Locks) at Sault Ste. Marie opened in 1855 and completed the last major link in the Great Lakes navigation system.

From the early 1820s until the onset of the Civil War in 1861, the Great Lakes region enjoyed an enormous

growth of population and trade. The excellent transportation system which the Lakes provided encouraged settlement. The population in the five states of the Old Northwest jumped from about 800,000 in 1820 to more than nine million by 1860, nearly one-third of the total population of the United States. Farm products, lumber, and coal went from west to east, while manufactured goods and immigrants moved in the opposite direction. In 1855, before Lake Superior iron and copper began to contribute to the totals, about four million tons of products worth $600 million moved through the Great Lakes, slightly more than the value of all the foreign trade of the United States. The number of ships in use was impressive as well. Steamboats first appeared on Lake Ontario in 1816 and on Lake Erie a year later. By 1836, the Lakes had 45 steam-powered vessels, mostly side-wheelers, and 217 sailing ships. Steamboats were used mainly for passenger traffic during this period. As late as 1860, there were 369 steam-driven ships on the Lakes versus 1,207 sailing vessels. The combined displacement of the Great Lakes fleet in 1860 was 463,000 gross tons, slightly less than one-tenth of the total for all American ships.

The first generation of U.S. lights on the Lakes came with the expansion of shipping and settlement. Seven lighthouses were built between 1818 and 1822, six on Lake Ontario and one on Lake Erie: the first was a tower (1818) attached to the mess-house at Fort Niagara, where the Niagara River flows into Lake Ontario; a light completed in 1819 at Presque Isle, at present-day Erie, Pennsylvania, on Lake Erie; the Galloo Island Light (1820); lights at Oswego (1822) and Rochester Harbor (1822); and two lights built in 1822 to mark the entrance to the Genesee River on Lake Ontario. Eight more lights went into service between 1825 and 1830: Big Sodus Bay Light (1825) on Lake Ontario; lights at Grand River or Fairport (1825), Buffalo (1828), Cleveland (1829), Otter Creek (1829) and Barcelona (1829), all on Lake Erie; the Fort Gratiot

Above left, the Rock Harbor Light, built in 1855, was closed by the Lighthouse Board in 1859, and again in 1879 after the copper and iron mines nearby were shut down. This photograph of the abandoned light was taken near the turn of the century. Photo courtesy State of Michigan Archives. Left, the Huron, Ohio, Light (1835) was another of the early lights on Lake Erie. This station was replaced by a modern structure in 1936. Photo courtesy The Great Lakes Historical Society. Top, an old photograph of the Michigan City Light, built in 1858. Photo courtesy Michigan City Historical Society. Above, a photograph of the Grand River or Fairport Light built on Lake Erie in 1825. Photo courtesy The Great Lakes Historical Society.

Top, Squaw Point Light (1897) on Little Bay De Noc helped guide vessels into Gladstone Harbor until the station burned down and the tower was removed by the Coast Guard. Photo courtesy Fred Ollhoff collection. Above, construction of the tower at the Grosse Point Lighthouse in 1873. Photo courtesy Evanston Historical Society. Above right, a day of pier fishing in the 1880s from the breakwater opposite Lake Park in Chicago. Photo courtesy Chicago Historical Society.

Lighthouse (1825), the first light on Lake Huron; and Bois Blanc Island Light (1829), also on Lake Huron. Construction boomed in the 1830s, with thirty-two new lights completed. The Chicago Harbor Light and the light at the entrance to the St. Joseph River, both completed in 1832, were the first on Lake Michigan. A total of 43 lights were in operation in 1840 including 17 on Lake Erie, 11 on Lake Michigan and 9 on Lake Ontario. Lake Huron was in fourth place with four lights, while Lake St. Clair and the Detroit River had one each. The first lightship on the Great Lakes was stationed at Waugoshance Shoal, just west of the Straits of Mackinac, in 1832 and served there each shipping season until 1851, when a lighthouse replaced it.

The Lighthouse Establishment added 33 new lights on the Great Lakes from 1841 to 1852, with Lake Michigan gaining 16. Lake Superior received its first six lighthouses, beginning with lights at Whitefish Point and Copper Harbor completed in 1849. The Great Lakes had a total of 76 operating lighthouses in 1852, when the Lighthouse Establishment operated 331 nationally. The relative importance of the individual lakes at that time can be seen from the distribution of lighthouses. Lake Michigan led with 27, Erie had 21, followed by Ontario with 9, Huron with 8, and Lake Superior with 6. Lake St. Clair and the Detroit River had a combined total of five. Two-thirds of the lighthouses were located at harbors or river entrances, while the rest marked islands, points, shoals, and reefs.

Lighthouse Administration Under the Fifth
Auditor of the Treasury, 1820-1852

Responsibility for the nation's aids to navigation passed to the Fifth Auditor of the Treasury in July, 1820, and remained there until 1852. The Fifth Auditor took over control of the Lighthouse Establishment and became the "General Superintendent of Lights," but this individual was also responsible for auditing the

books of a half dozen Federal agencies. Stephen Pleasonton was the Fifth Auditor in 1820 and he remained in that office until the entire system changed in 1852. He was a bookkeeper with little imagination and no maritime or engineering experience. Pleasonton was a hard-working, tight-fisted bureaucrat who emphasized economy to the detriment of other considerations, including safety. Under his administration, the nation's lighthouses deteriorated and their lighting systems became enormously inferior to those used by other maritime powers. Pleasonton's greatest mistake was his obstinate resistance to the use of the Fresnel lens during his long tenure in office.

In 1822 the French physicist Augustin Fresnel developed a radically different and vastly superior lens for use in lighthouses. It consisted of a series of glass prisms surrounding the light source in a lenticular (double convex) configuration, producing a lens which looked like a beehive. A central prism magnified the light while the prisms above and below refracted light to produce a single bright beam. The French and English lighthouse services quickly adopted the new lens, but Pleasonton dragged his feet. He had become a close friend of Captain Winslow Lewis and relied on Lewis for technical advice on lighting and other matters. Unfortunately, Lewis had become the supplier of illuminating equipment for virtually all U.S. lighthouses after 1812. The relationship between the two men was awkward if not corrupt. Lewis enjoyed the confidence of Pleasonton to such an extent that Lewis, without consulting anyone, once changed the characteristic of the Mobile Point Light on the Gulf of Mexico from a fixed, steady beam to a light that flashed. And as the years passed, Pleasonton found it increasingly more difficult to admit that Lewis's apparatus was of inferior quality, because it implied, that he, as the Fifth Auditor, had made an error.

The lighting system developed by Lewis had major defects which even the most objective of observers were able to spot. The reflectors, even when new, rarely had a parabolic form and were typically badly bent out of shape after several years' service. They were described by several investigators as ''spherical'' or having a ''wash basin design.'' The lightly silvered reflective surface did not last more than a few months because the standard cleanser in use was tripoli powder, an abrasive used on brass. Lewis fitted his apparatus with a tinted lens, which attracted smoke and dimmed the beam of light produced. One ship captain, commenting on southern lighthouses around 1850, observed, ''The lights on Hatteras, Lookout, Canaveral and Cape Florida, if not improved, had better be dispensed with, as the navigator is apt to run ashore looking for them.''

Congress received hundreds of complaints about U.S. lighthouses, most notably from George and Edmund Blunt's *American Coast Pilot,* the best guide available

Above, visitors at the Long Point Lighthouse, one of the earliest Canadian lights built on Lake Erie in 1843. Photo courtesy The Great Lakes Historical Society. Above right, the schooner Belle was purchased by the Lighthouse Board in 1863 to supply light stations on the Great Lakes. She remained in service until running aground in 1873. Photo courtesy The Great Lakes Historical Society. Right, lumber schooners were still sailing the Lakes in 1921. This unknown vessel is docked at the Coast Guard Station near Manistee, Michigan. Photo by J. Sherwin Murphy. Courtesy Chicago Historical Society.

to mariners. In June, 1838, Congress enacted a law dividing the Lighthouse Establishment into eight districts and appointed a naval officer for each district, with the specific task of inspecting all facilities and reviewing operations. The Great Lakes region was split into two districts, with the Detroit River entrance at Lake Erie as the boundary. Most of the inspecting officers severely criticized the lighting apparatus and the overall condition of the lighthouse buildings. They recommended that a qualified engineer supervise future construction projects. Congress eventually took several actions, but did not produce the real reforms which were needed. In 1838, Congress sent Commodore Matthew C. Perry to inspect European lighthouses and to buy a pair of Fresnel lenses. Perry did his job well and the lenses were installed in the twin towers of the Navesink, New Jersey, Light Station in 1841, with excellent results. As a ruse, Pleasonton praised the Fresnel lenses, but then made no attempt to use them elsewhere.

Congress continued to question the operation of the Lighthouse Establishment, but mainly in terms of costs and not performance. In February, 1842, the House Committee on Commerce was asked to review lighthouse expenditures since 1816 and to decide whether the management of the nation's lights should be transferred to the Topographical Bureau of the War Department. In May, the Committee issued a generally favorable report on Pleasonton's administration and argued against any transfer of the Lighthouse Establishment to another department. The Committee, did, however, recommend that permanent inspectors be appointed to make frequent assessments of lighthouse facilities. The Secretary of the Treasury appointed I.W.P. Lewis, a qualified civil engineer and the nephew of Winslow Lewis, as a special agent to inspect the lighthouses of New England, about one-third of the total maintained by the United States. The younger Lewis issued a scathing attack on his uncle's lighting system and by association, questioned the competence of

Winslow Lewis and Stephen Pleasonton. Despite the continuing barrage of complaints about the lights, Pleasonton could not be moved. In 1851, when the United States operated over three hundred major lights, only four Fresnel lenses were in service at three stations, all because of direct Congressional action.

After taking considerable pounding from wave after wave of complaints about the lighthouse system, Congress finally took action in March, 1851, when it required the Secretary of the Treasury to conduct a full-scale investigation of the Lighthouse Establishment. Secretary Thomas Corwin appointed a board consisting of two senior naval officers, two army engineers, a civilian scientist, and a junior naval officer who would serve as the board's secretary. The Board conducted an exhaustive study of the lighthouse system and on January 30, 1852 issued a 760 page report urging the introduction of the Fresnel lens and a complete reorganization of the administrative system, including the appointment of technically qualified inspectors for each lighthouse district. They specifically recommended that Congress create a nine-member Lighthouse Board to manage the system. The Secretary of the Treasury would serve as president *ex-officio.* Pleasonton tried to refute the report and urge the continuation of the *status quo,* but to no avail. On October 9, 1852, Congress enacted into law all of the board's recommendations, thus creating the administrative system which lasted until 1910. ☐

2 Expansion under the Lighthouse Board, 1852-1910

Throughout the second half of the nineteenth century, Great Lakes commerce continued to grow rapidly and remained a significant part of the national economy. Total shipments on the Lakes rose from about four million tons in 1852 to eighty million tons by 1910. The value of goods rose only fourfold in the interim, a reflection of falling prices and a major shift in the type of products shipped. Through the 1870s, lumber and grain accounted for three-quarters of total shipments, but by 1910, they made up less than one-tenth. Meantime, iron ore became the single most important cargo, accounting for half the 1910 tonnage, while coal made up another quarter. The iron mines of the Lake Superior region produced three-quarters of American ore in the early part of the twentieth century and virtually the entire output was moved by water to the major iron and steel plants at Gary, Detroit, Cleveland, and Buffalo.

The types of vessels used on the Lakes also changed considerably in the second half of the nineteenth century. Through the 1860s, sailing ships made up two-thirds of the tonnage on the Lakes. Once specialized iron and steel bulk carriers came on the scene in the 1880s, sailing ships became obsolete. The steam-driven, steel-hulled, propeller vessels quickly became the most economical means of moving the ore, coal and grain that dominated commercial trade. Steam tonnage pulled even with sailing tonnage in 1882, comprised two-thirds of the total by 1890 and three-quarters of the total ten years later. By the turn of the century, the basic design of the bulk carrier had emerged: a low-riding, long, slow-moving, steam-driven vessel with pilot house foreward, the power plant aft and little or no super-structure in between. During a remarkable building boom, Great Lakes shipyards finished 170 steel bulk carriers between 1905 and 1910. After the new ships were in service, 88 percent of the three million gross tons of vessels on the Great Lakes was steam-driven. The last commercial sailing ship in use on the Lakes was

Our Son, which foundered and sank in a storm on Lake Michigan in the fall of 1930.

By any reckoning, Great Lakes traffic was impressive. During the navigation season of 1888, for example, the port of Chicago had approximately 20,000 arrivals and departures of major vessels over eight months, compared to New York's 23,000 vessel movements spread over an entire year. In the course of the 1888 season, a total of 8,832 ships moved through the St. Mary's Falls Canal, while 31,404 vessels passed the Limekiln Crossing on the Detroit River, the outlet for Lakes Superior, Michigan, and Huron. The numbers at Limekiln Crossing represent an average of 140 vessels per day or six per hour. The Great Lakes fleet became relatively more important over time and by 1910, made up more than one-third of the tonnage of the entire American merchant fleet.

Lighthouse Design and Equipment

The network of aids to navigation on the Great Lakes expanded greatly under the Lighthouse Board. Between 1852 and 1860, the total jumped from 76 to 102. These figures reflect net gains because from the 1830s on, some lights were discontinued or destroyed. For example, the Maumee Bay Light fell victim to ice in 1856 and a gale force wind demolished the Big Sodus Bay beacon in 1857. Construction of new lights then slowed considerably, with only a dozen added in the 1860s, but in the next decade, forty-three new lights came into service, followed by more than a hundred new lights in the 1880s. The Lighthouse Service maintained an impressive set of aids to navigation by 1892: 219 major lights; 79 minor lights, mostly on pierheads; four lightships; 56 fog signals; and 404 buoys. Another construction boom took place in the 1890s and by the beginning of the twentieth century, the Great Lakes had 334 major lighted aids, 67 fog signals, and 563 buoys.

Lighthouse design changed gradually during the

Preceding page, a view inside a Fresnel lens at Rock of Ages Light. Above, in this sequence of photographs, the crew of a lifesaving station practices a Lyle Gun, or Line Throwing Gun, rescue. Here, they're loading the small cannon.

There were sixty-four lifesaving stations scattered throughout the Great Lakes. Another drill practiced routinely was capsizing the surfboat. Here, the crew of eight oarsmen row out to position in the harbor. At the helm, the keeper mans the rudder.

Firing the gun, a projectile with an attached "messenger line" is thrown to the target. The Lyle Gun was used when a shipwreck occurred within about 600 yards of the shore, and weather conditions made the use of the surfboat too dangerous.

Next, with oars secured, the crew stands on one side of the boat and reaches back to the grab the other side so that they can turn it over. A standard surfboat was 26 feet long and weighed between 700 and 1,000 pounds, depending on the wood used.

Once a second, heavier line was hooked up, the "breeches buoy" was sent out to the vessel by a pulley. It was a life buoy fitted with a canvas sling. Then, one by one, the endangered sailors were hauled to safety. Photos courtesy Jack Deo, Superior View.

Each surfboat was lashed with ropes so that the crew could pull themselves aboard. Photos courtesy Jack Deo, Superior View. Above right, the six-man crew of the Vermilion Lifesaving Station in 1915. Photo courtesy United States Coast Guard.

nineteenth century and several distinct types or styles emerged. The most common design until around 1870 consisted of a keeper's dwelling of wood, stone, or brick, with the light exhibited in a lantern built into the roof or mounted on an attached square tower. The focal plane of the light was typically between thirty and fifty feet above the water. By the 1870s, however, taller towers were needed, especially for coastal lights. The most common design was a conical brick tower, between eighty and one hundred feet tall, connected to the keeper's house by an enclosed passageway. The walls of the tower base had to be three or four feet thick in order to support the rest of the structure. The Lighthouse Board erected three identical skeletal iron light towers in 1861 at Whitefish Point and Manitou Island, both on Lake Superior, and at DeTour, at the southern entrance to the St. Mary's River. In the 1890s, they built a new generation of steel-framed skeletal lighthouses. Before the 1850s, there were also many different types of lanterns in use, including the "birdcage" lantern, but the eight or ten-sided polygonal design became nearly universal on the Lakes after 1850.

Lighthouse engineers, usually military personnel, developed a set of standard lighthouse designs during this period. This is why so many of the lights on the Lakes are virtually identical. Roughly a half dozen proven designs were used again and again, including standard designs for coastal beacons, river lights and

pierhead beacons. This practice saved vast sums of money since the engineers in the lighthouse districts avoided "reinventing the wheel" every two or three years. The lighthouses which are identical twins, triplets, quadruplets, etc., will be identified in later chapters. After the Civil War, the Lighthouse Board moved most harbor lights from the mainland onto newly built piers and breakwaters. Pier lights, while still manned, no longer included a residence, which typically remained on shore, so simple wooden or metal towers were sufficient. Few of these structures have survived from the nineteenth century because numerous pier extensions and the destructive effects of ice and storms shortened their lives.

The greatest problem faced by lighthouse engineers in the period 1870-1910 was the construction of light stations on isolated islands, reefs, and shoals. The Lighthouse Service had its own staff of engineers, but when a proposed lighthouse presented particularly difficult engineering problems, the U.S. Army Corps of Engineers would assign additional personnel to work on the project. They became expert in the design and construction of offshore lights resting on submarine crib structures, beginning with the Waugoshance Shoal Light (1851) and including significant lights at Spectacle Reef (1874), Stannard Rock (1882), and at the entrance to the Detroit River (1885). The culmination of these efforts was the completion of light stations at Rock of

Above, a portrait of Augustin Fresnel, the French physicist whose new lens dramatically changed the quality of lighting systems used in lighthouses throughout the world. Fresnel invented the lens that bears his name in 1822, just five years before his death, so he never knew the total impact that his invention would have on making navigation safer. Photo courtesy Musee de la Marine, Paris. Opposite page, top left, this Second Order Fresnel lens was removed from the Rock of Ages Light on Lake Superior after this photo was taken, but it is now on display at the Windigo Ranger Station on Isle Royale. Top right, a Third Order Fresnel lens at Presque Isle Light. Bottom left, a Third-and-a-half Order Fresnel lens from Sturgeon Point Light. Bottom right, a Fourth Order Fresnel lens from Harbor Beach Light. These three lights are located on Lake Huron.

Ages (1908) and at White Shoal (1910), projects which received considerable attention from the national engineering community. The histories of these major construction feats appear later in this volume.

Building a new lighthouse was sometimes deemed too hazardous or costly and the "floating lighthouse," or lightship, became a reasonably good substitute. After the light vessel stationed at Waugoshance Shoal from 1832 to 1851 was replaced by a lighthouse, no additional Federal lightships were on duty on the Great Lakes until October 1891, when three identical new vessels went into service: No. 55 at Simmon's Reef; No. 56 at White Shoal and No. 57 at Gray's Reef. The Craig Ship-building Company of Toledo, Ohio, built these wooden, steam-driven, propeller vessels, each 102 feet long and equipped with two lanterns, one mounted on the mainmast and the other on the foremast. They were the first American lightships with engines, but these must have been small power plants because the tender *Dahlia* towed the three vessels to their stations on Lake Michigan, just west of the Straits of Mackinac. Each was equipped with a five-ton anchor to prevent the vessel from losing its moorings in bad weather, a common problem for lightships of that era. During their first season, all three left their stations by November 20th, before the end of the shipping season, ignoring explicit orders in the process. The Lighthouse Service dismissed the officers and crews for dereliction of duty. By 1899, seven additional lightships were on duty, at Eleven Foot Shoal (Lake Michigan), Poe Reef (Lake Huron), Grosse Pointe (Detroit River), two at Limekiln Crossing (Detroit River), Ballard Reef (Detroit River), and at Bar Point (Detroit River). New lighthouses and buoys eventually reduced the need for lightships. Still, as late as 1925, the Lighthouse Service had eleven vessels on duty on the Lakes, although not at the same stations as a quarter of a century earlier.

With the Lighthouse Board making all technical decisions about equipment beginning in 1852 after

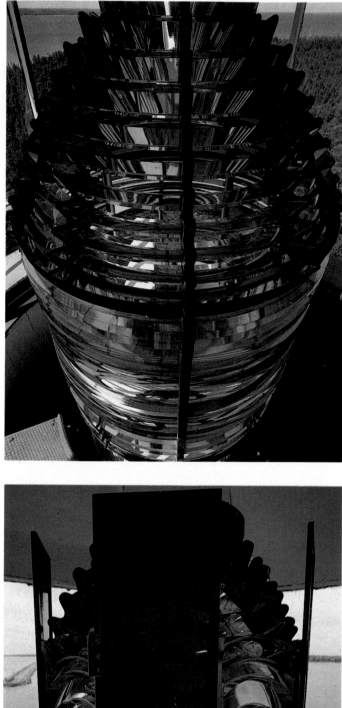

25

Above, the last commercial sailing ship on the Great Lakes was the Our Son, *shown in this photograph in 1922, eight years before she foundered and sank in a storm on Lake Michigan in the fall of 1930. Above right, this 1884 engraving by artists Schell and Graham shows a view of the city from the top the Chicago Harbor Light at the mouth of the river. Both photos courtesy Chicago Historical Society.*

Stephen Pleasonton's departure, the Fresnel lens was quickly put into service nationally. The Board refitted seventy-five lights on the Lakes between 1854 and 1857 and placed Fresnel lenses into all new lighthouses as well. Three French lens builders supplied the lenticular apparatus in those years: Henri Le Paute; Barbier & Fenestre; and Sautter, Lemonier & Cie. The new lenses typically doubled the five to seven mile range of the existing lights to a range of ten to fourteen miles. The new lenses also reduced fuel consumption to about one quarter of what had been common with the old Winslow Lewis system.

Fresnel lenses are classified into six "orders," based on the focal length of the lens, but seven sizes exist because there is a Third-and-a-half Order lens. The First Order lens, the largest, has a focal length of 36 inches, giving the lens a diameter of six feet. It stands nearly twelve feet tall. This size was used only in the largest seacoast lights. The smallest, the Sixth Order

lens, has a focal length of 5.9 inches, a diameter of just under one foot and height of about two feet, and was normally used only on pier or breakwater lights. The Second Order lens was the largest ever used on the Great Lakes, and only at five major stations. The 1858 *Light List* gives the distribution of Fresnel lenses on the Lakes: five Third Order; 40 Fourth Order; 14 Fifth Order; and 41 Sixth Order. In 1892, when the Lighthouse Service maintained 219 lights on the Lakes, only three were Second Order and 23 were Third Order. The Fresnel lenses were slowly replaced beginning in the 1920s, but there are still about one hundred in use.

New lamps also went into service along with the Fresnel lenses, replacing the older Argand type. Several inventors developed lamps, but the most common bore the names of Funck, Carcel, and Le Paute. George C. Meade, who led the Union forces at Gettysburg, worked for the Lighthouse Service as a young engineer and designed a popular lamp around 1853. The lamps shared similar designs, utilizing from one to four concentric wicks, depending on the amount of light required. Illuminants changed in the late 1850s as well, largely because of the extinction of the sperm whale. The cost of whale oil increased sharply from 50 cents per gallon in 1840 to $2.25 per gallon by 1855, forcing the Lighthouse Board to look for a cheaper alternative. The Board asked Professors Morfit and Alexander of the University of Maryland to study the problem and they conducted a set of experiments utilizing several grades and combinations of whale, shark, fish, seal, colza, lard, and mineral oils, producing an impressive report on these illuminants. The two scientists recommended the use of colza (rapeseed) oil and the Board tried it in the late 1850s, but the supply was limited because its source, wild cabbages, were not commonly grown in the United States. One of the Board's scientists, Professor Joseph Henry of the Smithsonian Institution, resumed tests with lard oil and discovered that it burned well if preheated. By the late 1860s, lard oil had become the most common illuminant in use, particularly at the larger lighthouses. Throughout the 1870s, the Lighthouse Service consumed an average of 100,000 gallons of lard oil annually. Although earlier efforts to use natural gas at the Portland (Barcelona) Light on Lake Erie in the 1830s had failed, the Board tried again in 1866, but failed again.

A new set of trials with mineral oil (kerosene) proved successful and in 1877, the Board began to convert lights to the new fuel. By 1885, kerosene became the dominant illuminant, in part because of the invention of the incandescent oil vapor lamp. In this device, the kerosene turned into a vapor after striking the hot walls of a vaporizing chamber and then passed into a mantle where it burned with a bright light. Compared to the lamps of the wick type, the incandescent oil vapor lamp generated a brighter light with no increase in fuel

Above, a rare photograph of a steam whistle in action. This is one of two ten-inch steam whistles that replaced a pair of steam sirens used as fog signals at the Grosse Point Light in 1892. Above right, a close-up of the whistle. The sound deflector measured nine feet in diameter. These signals were removed in 1922. Both photos courtesy Evanston Historical Society. Opposite page, above right, an early 1890s photograph of the Northwestern University Lifesaving Station. Photo courtesy Evanston Historical Society. Right, two members of the lifesaving station at Eagle Harbor on Lake Superior. Photo courtesy Mrs. Janice Gerred.

consumption. By 1889, when the new lamp was almost universally used, the Lighthouse Service burned more than 330,000 gallons of kerosene. The Board conducted some experiments with electric arc lamps as early as 1886, but this new energy source was insignificant until the early part of the twentieth century.

Fog signals also became an important element in the system of aids to navigation. In 1858, the Lakes had only four fog signals in service, all bells struck by machinery. In the 1850s, the Lighthouse Board experimented with a variety of fog signal devices, including cannons, bells driven by clockwork mechanisms, giant whistles or trumpets with compressed air causing a reed to vibrate, steam sirens, and finally, steam locomotive whistles. The last device proved to be the most successful of all. The Great Lakes boasted 56 fog signals by 1892, including 16 bells driven by clockwork and 40 steam whistles, mostly of the locomotive-type, which apparently came in two standard sizes, six-inch and ten-inch. Steam whistles were used on the Lakes well into the 1920s.

The Lighthouse Board initially chartered vessels to maintain and supply its lights on the Great Lakes, but from 1857 on, they operated from three to five of their own lighthouse tenders. In that year, the Board purchased two schooners and renamed them *Lamplighter* and *Watchful*. The schooner *Dream* replaced *Lamplighter* in 1862 and *Watchful* was sold in 1867. The Board purchased the schooner *Belle* in 1863 and she remained in service until running aground in 1873. The first steam-powered propeller to serve as a tender was the *Haze*, which remained in service from 1867 until 1905. Another steamer, the *Warrington*, helped with the construction of the Spectacle Reef Light beginning in 1871 and remained in service through 1911. The *Dahlia*, a propeller steamer, was the first tender specifically designed for the Great Lakes and was also the first with an iron hull. She served from 1874 until 1909 and marked the beginning of the practice of naming tenders

after plants, flowers and trees. Other lighthouse tenders of this era included the *Lotus* (1880-1901), the *Marigold* (1890-1946), and the *Amaranth* (1891-1946).

ADMINISTRATION UNDER THE LIGHTHOUSE BOARD

The first Lighthouse Board began its work on October 9, 1852, concentrating most of its efforts on upgrading existing facilities and building new ones. For purposes of administration, the Great Lakes remained divided into two districts with the Detroit River entrance into Lake Erie as the boundary. The Civil War severely disrupted the Lighthouse Service because the engineers and inspectors, all military men, were reassigned to other duties, while at the same time, the Confederates captured or destroyed at least 160 lights. A Senate Bill introduced in 1862 proposed transferring the Lighthouse Service to the Navy Department, but the bill failed. The Secretary of the Navy then launched a second campaign between 1882 and 1885 to bring the Lighthouse Service under his control, but this effort also failed. Congress consistently took the position that lighthouses existed to serve commerce and therefore the Lighthouse Service belonged in the Treasury Department, which at that time had full responsibility for commerce.

The Lighthouse Board brought order to the nation's system of aids to navigation. Beginning in 1838, during Stephen Pleasonton's administration, the Lighthouse

Above, the lighthouse tender Aspen *makes a stop at the Spectacle Reef Light in Lake Huron. Put into service at Toledo, Ohio in 1906, the* Aspen *was finally laid up in 1948. The ship had a steel hull and she measured 126 feet in length. Photo courtesy Dossin Great Lakes Museum. Above right, the iron-hulled* Marigold *was launched at Wyandotte, Michigan in 1890. After she was taken out of service in 1946, her hull was used to build the dredge* Miss Mudhen II. *Photo courtesy The Great Lakes Historical Society. Right, a 1916 photo of the crew from the* Marigold. *Photo courtesy Dossin Great Lakes Museum.*

Establishment periodically published a *List of Lighthouses, Beacons and Floating Lights of the United States,* but the lists were often inaccurate and generally of little value to mariners because the lights were arranged chronologically, by state. The Lighthouse Board issued a vastly improved *Light List* in 1852 and beginning in 1869, revised it annually. The Board initially established a central supply depot at Staten Island, New York, but soon built storage facilities in each district. In Detroit, they established a "buoy and supply depot" on the grounds of the U.S. Marine Hospital on East Jefferson Avenue in 1869 and then built a handsome brick warehouse at the foot of Mt. Elliott Avenue near the Detroit River, completing the building in 1874. Initially, each of the twelve lighthouse districts had either an Army or a Navy officer who served as inspector and superintendent. By the 1860s, the usual practice was to have an Army officer serve as engineer and a Navy officer as inspector in each district. The United States

was divided into twelve districts in 1852, but Congress increased the number to sixteen in 1886. Under the new organizational scheme, which remained in effect until 1910, the Great Lakes were split into three light districts: the Ninth, consisting of Lake Michigan; the Tenth, encompassing Lake Ontario and Lake Erie; and the Eleventh, which included the Detroit River, Lake St. Clair, the St. Clair River, Lake Huron, the St. Mary's River and Lake Superior.

Throughout their years of jurisdiction, the Lighthouse Board continued to make significant improvements in the lighthouse system of the United States. The Board turned what had been a distinctly inferior system in 1852 into the most advanced as well as the largest in the world by the early twentieth century. The number of major lights jumped from 332 in 1850 to nearly 4,000 by 1910, while the grand total of all aids to navigation increased from about 1,500 to nearly 12,000 over the same six decades. Administering such a large system became increasingly difficult under the cumbersome nine-member Lighthouse Board. The legislation of February 14, 1903, which created the new Department of Commerce and Labor, also transferred the Lighthouse Service from Treasury to the new department. The first Secretary of Commerce and Labor, Oscar Strauss, and his successor, Charles Nagel, argued that the lighthouse system suffered from the lack of a single executive with clearcut authority and from the division of authority that

existed within each district between the two military
officers. Congress eliminated the Lighthouse Board on
June 17, 1910, and created the Bureau of Lighthouses in
its place. The new bureau was staffed almost entirely by
civilians, thus ending, at least for a time, the military
influence on the lighthouse system. The three river
districts, two on the Mississippi and one on the Ohio,
were placed under the control of officers from the Army
Corps of Engineers.

LIFESAVING ON THE GREAT LAKES

As the volume of traffic grew on the Lakes, greater
numbers of ships ran aground or sank in bad weather,
often with great loss of life, but the Federal government
moved slowly to provide life-saving services to mariners.
In 1807, the Massachusetts Humane Society established
the first permanent lifeboat station in the United States
at Cohasset, Massachusetts. Congress did not become
actively involved in lifesaving until 1837, and then only
through periodic, haphazard spending for lifeboat
stations and equipment. The first appropriation for
Great Lakes lifesaving came in August, 1854, when
Congress provided $12,500 for the purchase of lifeboats
for twenty-five stations, but there was no provision for
hiring full-time crews to man them. The resulting
system was unreliable and often inept, but remained in
place until the disastrous winter of 1870-71, when 214

men died on the Great Lakes. Responding to public outrage, Congress appropriated $200,000 in April, 1871, to build new lifesaving stations nationwide, and to hire paid crews. The construction of new facilities did not begin in earnest until 1874 and new stations on the Great Lakes did not open until September, 1876.

The first generation of lifesaving stations included 27 on the Lakes: Ontario, Huron, and Superior had four apiece; Lake Erie received five; and Lake Michigan led the region with ten. The stations on Lake Huron were all built near existing lighthouses, but this was not the usual practice on the other lakes, where harbor entrances were the most common location. Those built on Lake Superior were erected at desolate locations on the eastern end of the lake, at Vermillion Point, Crisp's Point, the Two Hearted River, and the Sucker River. On June 18, 1878, Congress made the Life-Saving Service a separate agency within the Treasury Department and President Rutherford B. Hayes

Above left, the Amaranth *was built in 1891 and served until 1946. After World War II, she was sold to a freight company that operated between Michigan and Wisconsin and was renamed the* South Wind. *Photo courtesy Great Lakes Graphics. Left, the tender* Crocus *is being repainted in dry dock at Buffalo, New York in July, 1917. Photo courtesy Dossin Great Lakes Museum. Above, the* Dahlia *tended the light stations on the Great Lakes from 1874 until 1909. She was the first tender specifically designed for the Great Lakes, and was also the first tender built with an iron hull. Photo courtesy The Great Lakes Historical Society.*

In the last quarter of the nineteenth century, the Lighthouse Board began to change the characteristic of many of its lights on the Great Lakes. The reason they changed a fixed light to one that flashed was because there were often several light stations in the same vicinity. With all of them exhibiting a fixed light, it was difficult for the mariners to tell them apart. In general, the flash was produced by installing panels mounted in a frame work that revolved around the outside of the lens. The movement of this frame was timed and powered by a clockwork mechanism which was driven by weights. These clocks would then have to be rewound at various intervals of 8 to 24 hours. Above are two of these mechanisms. At left is the apparatus at the Split Rock Light. At right, the clockwork mechanism from Grosse Point Light. Opposite, above, is a blueprint showing one of the standard lighthouse designs that were used in the nineteeth century. At right, a detail of a light tower. Both photos courtesy State of Michigan archives.

appointed Sumner Kimball as General Superintendent. Kimball held the post until 1915, when the Life-Saving Service became part of the new United States Coast Guard. By 1893, 47 lifesaving stations served the lakes and 60 were in service by 1900, but only four new ones opened after the turn of the century.

Lifesaving stations had a keeper responsible for the buildings, equipment, and the crew of six or eight "surfmen." All personnel lived at the station during the shipping season, but were paid only for that time as well. The crew spent most of its time maintaining watches in lookout towers or patrolling miles of beaches looking for vessels in distress. The crew regularly performed drills with various signal devices, the surfboats, and with the station's "beach apparatus," transported on the "beach cart," usually pulled by the crew. When a shipwreck occurred within about 600 yards of the shore, a typical situation, and weather conditions made the use of the surfboat too dangerous, the "beach apparatus" came

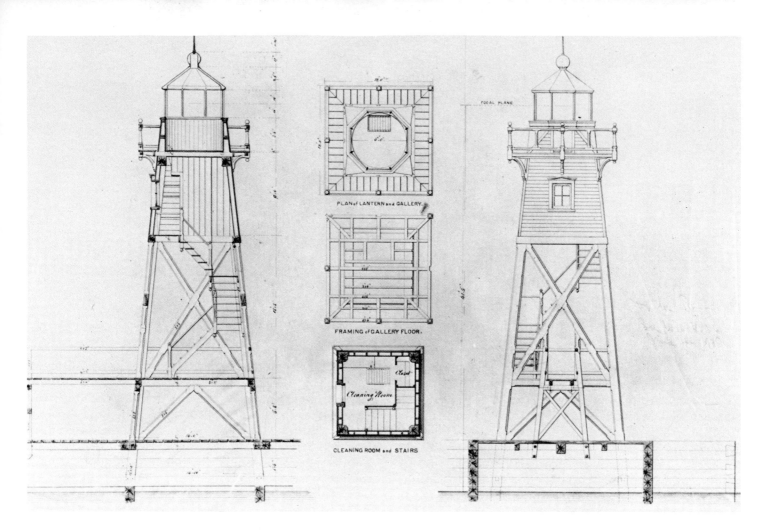

PLAN of LANTERN and GALLERY.

FRAMING of GALLERY FLOOR.

CLEANING ROOM and STAIRS

into play. The key piece of equipment was the Lyle Gun, or the Line Throwing Gun, invented by David A. Lyle. This small cannon fired a projectile with an attached "messenger line," which would reach a vessel wrecked close-by. Once the first line was firmly attached, a heavy duty line was extended out to the vessel and then solidly tied to a mast or other high point on the ship. After this was done, a rescue device known as a "breeches buoy" would be moved out to the vessel on a pulley hanging from the line. It was a round life buoy with a canvas sling resembling a pair of breeches suspended from the buoy. One at a time, the endangered mariners would climb into this device and be safely moved to the shore, suspended high above the heaviest seas. Of course, the vessel in distress had to remain afloat.

The Lyle Gun was used in hundreds of successful rescues on the Lakes, involving several thousand lives. The beach apparatus gradually went out of service, but a few Lyle Guns were still in use in the late 1950s. The motorized lifeboat, developed in 1889 in Marquette, Michigan, by Keeper Henry Cleary and the Lake Shore Engine Works, became the mainstay of the Life-Saving Service and the U.S. Coast Guard in the early twentieth century. With the addition of radios, the new lifeboats greatly increased the speed and range of lifesaving operations. The rescues continue and the rescuers are no less heroic than in the past, but much of the excitement and romance are gone. □

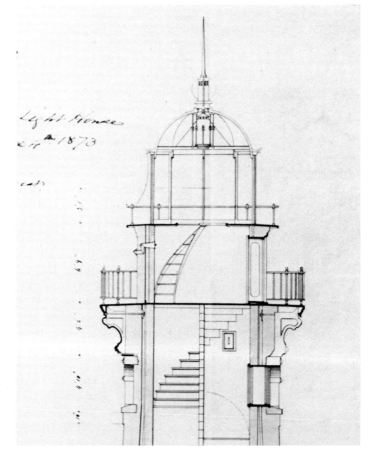

3 Automation and Abandonment in the Twentieth Century

Commerce on the Great Lakes has continued to grow over the past seventy-five years, making this navigation system as important as ever. The total shipments of about eighty million tons of products in 1910 more than doubled in three decades, reaching 169 million tons in 1941. By the late 1960s and early 1970s, shipments averaged more than 200 million tons per annum and iron ore has consistently made up half the total. Some shrinking of traffic and tonnage occurred in the late 1970s, mostly because of the decline in American steel production. The importance of the Great Lakes can be seen through one comparison: from the opening of the Panama Canal in 1915 to the present, the total tonnage passing through the locks at Sault Ste. Marie has been greater than the tonnage passing through the Panama Canal and Suez Canal combined. This is even more impressive when we take into account the short shipping season of Lake Superior.

The bulk carriers which move this freight have grown progressively larger over the years. A typical bulk carrier at the beginning of the twentieth century was about 400 feet long with a carrying capacity of 5,000 tons. The building of progressively larger "super carriers" has been a major feature of the Lakes trade, with shipbuilders limited only by the size of the locks and depths of the channels on the system. Four freighters of 8,000 ton capacity went into service in 1900, followed by a 12,000 ton ship in 1904, and two years later by the *J. Pierpont Morgan,* a 600-footer with a 13,000 ton capacity. Over the following three decades, freighters ranging from 500 to 600 feet in length became the standard. With the wartime growth of steel output, a new generation of freighters came into service in 1943, measuring 640 feet long, with a 24-foot draft and a capacity of 18,600 tons. The past few decades have brought even further escalation of size. The ore carrier *Roger Blough,* put into service in 1972, was 858 feet long and carried 45,000 tons of iron pellets. In the following two years, the *Stewart J. Cort* and the *Presque Isle* became the first 1,000-foot freighters on the Lakes. Only a handful of the new generation of super-sized, bulk carriers went into service in the 1970s, but the 1,000-footers, which are 105 feet wide and have a capacity pushing 60,000 tons, are clearly the wave of the future. There's even talk of building 1500-footers! The size of such enormous freighters is almost beyond comprehension. By comparison, the Sears Tower in Chicago, currently the world's tallest building, stands 1,454 feet high.

The size of vessels on the Great Lakes, however, has been limited by the dimensions of four bottlenecks: the locks at Sault Ste. Marie; the St. Mary's River; the shipping channel dredged through Lake St. Clair; and the Detroit River channels near Lake Erie. The first two locks at Sault Ste. Marie, opened in 1855, measured 350 feet long, 70 feet wide, and nine feet deep. They were replaced by the Weitzel Lock (1881), with dimensions of 515 by 60 by 11.4 feet. An additional lock, named after General O. M. Poe, opened in 1896 and measured 800 by 100 by 18 feet. Two additional locks, the Davis (1914) and the Sabin (1919) had identical dimensions of 1,350 by 80 by 25 feet. These early locks were built on the assumption that they would handle four vessels at a time. In 1942, the new MacArthur Lock opened, with dimensions of 800 by 80 by 31 feet, a significant increase in size, especially depth, over the Weitzel Lock which it replaced. Finally, in response to the projected new superfreighters, the Army Corps of Engineers opened a new Poe Lock in 1969 on the site of the first Poe Lock. It stands 1,200 feet long, 110 feet wide, and 31 feet deep. The Corps of Engineers is now planning to replace the Davis and Sabin Locks with a new one roughly the same size as the new Poe Lock. When vessels larger than canoes began using the Great Lakes in the 1820s, the rock reefs at the Limekiln Crossing on the lower Detroit River and the St. Clair Flats mud shoals offered channels only six feet deep. Over the years those channels and the St. Mary's River below the Soo Locks

have been dredged or blasted to accommodate the ever-increasing size of the larger vessels. Today, the official depth provided for navigation is 27½ feet (low water datum) in all channels.

The system of aids to navigation on the Lakes continued to expand in the early part of this century. Nationwide, the Lighthouse Bureau completed a grand total of 38 major new light stations or light systems between 1910 and 1925, with 13 of these, or one-third of the total, located on the Great Lakes. In 1925, the Lakes system had 433 major lights, including pier lights, 10 lightships, 129 fog signals, and about 1,000 buoys, producing a grand total of 1,771 aids. Only 160 light stations actually had resident keepers, so the overwhelming majority of aids were unmanned. Virtually all the lighthouses in use today were in place by 1925. Most later construction involved rebuilding or relocating existing lighthouses or replacing lightships with permanent light stations. With the development of radio as a major element in the Lakes navigation system, the Bureau of Lighthouses began a long process of eliminating lighthouses. The total number of aids kept growing and stands at over 2,500 today, but almost all the new ones added since the 1920s were buoys.

Lightships practically disappeared by the Second World War. In 1925, ten served at regular duty stations and the Lighthouse Service kept one relief vessel as a spare. Milwaukee, Peshtigo Reef (Wisconsin), the Corsica Shoals in Lake Huron, and Lake St. Clair each had lightships, with the remaining six located on reefs and shoals in the vicinity of the Straits of Mackinac. Lightship No. 56 (1891) was still in service, as was No. 60 (1893), a wooden sailing vessel. By 1940, only one remained. The *Relief,* built in 1921, was renamed the *Huron* in 1935 and stationed at the Corsica Shoals at the southern end of Lake Huron, where she served until decommissioned in 1970. Fortunately, this vessel is preserved on the St. Clair River in Port Huron. The Bureau of Lighthouses enlarged its fleet of lighthouse tenders as well. In 1925, the Great Lakes had seven tenders, all steam propellers built between 1892 and 1906, ranging from 93 to 169 feet in length. By 1940, nine tenders were on duty, including the sister ships *Hollyhock* (1937) and *Walnut* (1939), each 175 feet long.

Most of the fog signals used on the Lakes at the turn of the century were steam-driven locomotive whistles. Their great drawback was the length of time needed to raise a full head of steam, usually about an hour. Often, by the time the fog whistle began to sound, the fog had disappeared or the unfortunate vessel had run aground or sank. Gradually, internal combustion engines driving air compressors became the most common type of fog signal before radio came into widespread use in the late 1920s. In 1925, when 127 fog signals served the Great Lakes, 104 were driven by gasoline or diesel engines. The rest included thirteen clock-driven devices, one that

was worked by hand, eight powered by electricity, and one radio signal, newly installed.

THE BUREAU OF LIGHTHOUSES AND MODERNIZATION, 1910-1939

On June 17, 1910, Congress eliminated the nine-member Lighthouse Board established in 1852, and placed the Lighthouse Service under the new Bureau of Lighthouses, part of the Department of Commerce. Under the new administrative arrangement, the military influence on the system was virtually eliminated, although an officer from the Army Corps of Engineers continued to serve as a consultant in each district. The first Commissioner of Lighthouses under the new system was George R. Putnam, a civil engineer, surveyor, and civil servant with a long and distinguished record. Arthur Conover served briefly as Deputy Commissioner, followed by John S. Conway. The Chief Construction Engineer from 1912 on was H. B. Bowerman, while Edward C. Gillette held the post of Superintendent of Naval Construction starting in 1914. Putnam served for 25 years, Gillette for 21, Bowerman also for 21, and Conway for 20 years, producing stability at the top almost to the time the Lighthouse Bureau went out of existence in 1939.

During the Bureau of Lighthouses era, which was really the George R. Putnam era, the Lighthouse

Preceding page, the modern skyline of Chicago overshadows the "new" harbor lighthouse originally built in 1893, but moved to its present location in 1917. Opposite, construction of the light station at Split Rock on Lake Superior was an engineering triumph because the men and materials had to be hoisted up the face of the cliff some 124 feet in height. This photograph shows the work in progress during the winter of 1908-09. Photo courtesy Minnesota Historical Society. Above left, construction of the concrete casing at the Grosse Point Light on Lake Michigan in 1914. Above, a night photograph of Grosse Point that was taken in 1946. Both photos courtesy Evanston Historical Society.

Service continued to expand, while at the same time, became the worldwide leader in technological innovation. The system of 11,661 aids in 1910 included 1,397 major lights and 54 lightships nationwide. By 1939, the total number had more than doubled to 29,606 aids. Because of the improved accounting and inspection systems introduced in 1912, the Bureau was able to manage a larger system more effectively. The trend toward the use of automatic equipment at light stations accelerated during these years. Despite the substantial growth in the system of aids between 1910 and 1939, the number of employees fell from 5,832 to about 5,200 over these years. In its 1936 annual report, the Bureau pointed out with great pride that "the Lighthouse Service is the most extensively decentralized agency of the Federal government," with less than one percent of its personnel in Washington, D.C.

Under Putnam's leadership, the Bureau remained at the forefront of technical and scientific advances.

Opposite page, top left, a large crane sets the light tower in place at Sheboygan, Wisconsin in 1915. The lighthouse was brought to this location after being constructed elsewhere. Photo courtesy Dossin Great Lakes Museum. Opposite, top right, construction of the William Livingstone Memorial Light on Belle Isle in 1929. Photo courtesy Dossin Great Lakes Museum. At left, the lightship Buffalo *served on Lake Erie until it foundered in a storm in 1913. Photo courtesy United States Coast Guard. Top left, the lightship* Poe *was stationed on Lake Huron near Poe Reef in the early decades of the twentieth century. Photo courtesy Dossin Great Lakes Museum. Left, the lightship* St. Clair. *Photo courtesy State of Michigan Archives. Above, the* Bar Point *was one of the first lightships to serve on the Lakes. Photo courtesy Dossin Great Lakes Museum.*

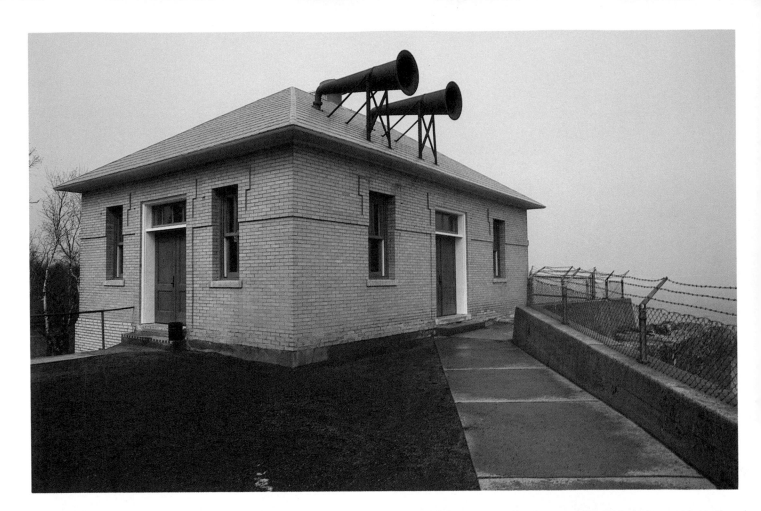

42

Electric power and one of its offspring, radio, substantially changed the system of navigational aids. Although experiments with electricity as a light source began in the 1880s, no widespread use took place until the twentieth century. The Lighthouse Board installed electric arc lights in the torch of the Statue of Liberty in New York Harbor in 1886, the first use of electricity for lighthouse purposes in the United States. In 1888, the Board placed six electric buoys in the Gedney Channel into New York Harbor, but replaced them in 1903 with cheaper gas buoys. The Lighthouse Service installed electric incandescent lamps in the Sandy Hook Lighthouse in New York Harbor in 1889 and on Lightship No. 51 in 1892. The experiments continued with the installation of an electric arc lamp and a complete generating station at the Navesink, New Jersey, lighthouse in 1898.

Electricity did not come into general use in lighthouses until the late 1920s, when the electric grid system became widespread and inexpensive portable generators were available. As early as 1925, electricity provided light for 68 major and 45 minor light stations on the Lakes, or roughly one-quarter of the total. The conversion to electricity was even faster after 1925 and by the early 1940s was nearly complete on the Great Lakes. In 1916, the Lighthouse Service introduced a device which automatically changed burned-out incandescent bulbs, moving a fresh one into place. Automatic timing mechanisms to turn the lights on and off were also introduced, making automation feasible at many stations. By the 1930s, the Service had also developed fog signals driven by electricity and activated by remote control. Together, these inventions rendered most manned lighthouses obsolete.

The use of electricity also encouraged and facilitated the first major changes in lenses since the 1850s. During the last half of the nineteenth century, virtually all lighthouse lenses were the Fresnel type, imported from the French. Starting in 1912, the Lighthouse Bureau

Opposite, above, the fog signal building at the Split Rock Light on Lake Superior. Opposite, below left, a pair of foghorns and the back-up light at Chicago Harbor. Opposite, below right, a fog horn at Whitefish Point Light on Lake Superior. Above left, the Escanaba Light, built in 1938, is typical of the large number of modern, automated lights put into service by the Coast Guard over the past 50 years. Above, the light station at Outer Island in the Apostle Islands in Lake Superior is an example of even more sophisticated technology. This lighthouse is now powered by solar energy. The three solar panels are attached to the rail.

Top, the Great Lakes shipping fleet numbers several hundred vessels, and the freighter Black Bay *is a part of the fleet operated by the Canada Steamship Lines. She is 730 feet long and has a maximum carrying capacity of 27,350 tons. Above, Gravelly Shoal Light, located in Lake Huron's Saginaw Bay, is a radio beacon that was put into service in 1939. At right are the four U.S. locks and one Canadian lock at Sault Ste. Marie. Vessels traveling the full length of the Great Lakes and St. Lawrence Seaway, from Lake Superior to the Atlantic Ocean, must pass through a total of sixteen locks. A new lock proposed for the Soo would enable the lock system there to accommodate about 138 million tons of cargo annually.*

encouraged American glass manufacturers to produce lenses and several alternatives to the Fresnel apparatus came into use. By the 1920s, new lighthouse installations rarely used Fresnel lenses, although those already in place remained in service. Lenses similar to those used in locomotives were often adapted for range lights after the turn of the century. The self-contained lens-lantern evolved, with an electric light encased in a glass lens, which also served as the lantern. This design could be exposed to the elements and was particularly suited for use on piers, breakwaters and posts. By 1925, the Great Lakes already had 48 locomotive-type lenses and 228 lens-lanterns in service, compared to 193 Fresnel lenses.

Radio beacons, which in effect served as noiseless fog signals and lightless aids, came into general use in the late 1920s. The Lighthouse Service first experimented with radio beacons in 1917 and in 1921 installed its first permanent radio beacons on two lightships and at one lighthouse. In 1925, there were thirteen in use nationwide, including one on the Lake Huron lightship, put into service there on June 12, 1925, the first of its kind on the Great Lakes. When several fixed radio beacons were in use, vessels could easily determine their position by taking bearings on several signals. By 1942, the Lakes had a total of sixty-five radio beacons, including seven in Canada. The Lighthouse Service also initiated the broadcasting of "Notices to Mariners Regarding Aids to Navigation" by radiotelephone from Sault Ste. Marie in 1937 and five additional stations went into service the following year. With this system, the Lighthouse Service could instantly notify mariners about malfunctioning lights, lost buoys, and other hazards. They also passed on information on weather and ice conditions, water levels, and related news.

THE COAST GUARD ERA, 1939 TO THE PRESENT

With the implementation of President Franklin D. Roosevelt's Reorganization Act of 1939, the Bureau of

The fascination that many people have about lighthouses is demonstrated in many ways. Top, this lighthouse was made for the birds at Sherwood Point Light, Wisconsin. Above, Jayne Clement of Kalamazoo, Michigan has been creating counted cross stitch designs of many Great Lakes lighthouses for the past three years. This is one of her designs. Opposite are four examples of lighthouses that have been abandoned. Top left, Turtle Island Light near Toledo, Ohio. Top right, the old lighthouse at Indiana Harbor near Chicago. Bottom left, Long Tail Point Light in Green Bay on Lake Michigan. Bottom right, Minnesota Point Light near Duluth on Lake Superior.

Lighthouses went out of existence on July 7, and the Lighthouse Service was transferred to the United States Coast Guard in the Treasury Department. Lighthouse Service personnel were allowed to retain their civilian status or join the Coast Guard, with their previous salaries unchanged. About half of the 5,200 employees chose each option. The 1939 Act was the last of several efforts to bring all Federal maritime activities under one roof, thus eliminating duplication and improving efficiency. President William Howard Taft had proposed a merger of the Lighthouse, Life-Saving, and Revenue Cutter services in 1912, but Congress would not approve. Instead, Congress merged the last two agencies in 1915 to form the Coast Guard and left the Lighthouse Service independent. The Revenue-Cutter Service, established in 1790 to enforce customs regulations, also had responsibility for aiding vessels in distress after 1836. The 1939 merger simply implemented Taft's earlier proposal.

The Coast Guard was absorbed into the United States Navy on November 1, 1941 and remained there until January 1, 1946, when it returned to the Treasury Department. The Great Lakes became the Ninth Coast Guard District and kept this designation even after the agency became part of the new Department of Transportation in 1967. Since 1946, the Coast Guard has maintained between 2,000 and 2,500 aids to navigation on the Great Lakes, usually serviced by seven major tenders. The increased use of electronic aids to navigation, ranging from radar to satellites, has greatly reduced the need for lighthouses. In the past four decades, the Coast Guard has taken scores of lights entirely out of service and has automated the rest. By 1965, only eighty lights plus the *Huron* lightship were still manned. The process was completed in the Fall of 1983 when the last manned light stations on the Lakes, at Point Betsie, Michigan and at Sherwood Point, Wisconsin, were fully automated. Thus, the romantic era of these lights has drawn to a close. □

4 THE LIGHTKEEPERS' WORK AND FAMILY LIFE

Beginning with George Worthylake's appointment as keeper of the lighthouse on Little Brewster Island in Boston Harbor in 1716, the lighthouse keeper's job was often seen as a plum ripe for the politician's picking. During the early years of Federal control over lighthouses, the President was responsible for appointing keepers. Presidents Washington, Adams, and Jefferson showed great interest in the lighthouse system and personally approved all of the appointments and dismissals of keepers. When Jared Hand asked in 1808 to be appointed keeper of the Montauk Point Light on Long Island, to succeed his father, President Thomas Jefferson emphatically ruled against making the position hereditary. Beginning with President James Madison, the local collector of customs usually nominated keepers, but the Secretary of the Treasury still made the appointments, a practice which continued until 1896, when keepers finally came under the Federal civil service system. Throughout most of the nineteenth century, however, the collectors of customs, political appointees themselves, often nominated keepers to repay political favors and, subsequently, often undermined the best efforts of the Fifth Auditor of the Treasury and the Lighthouse Board to develop a merit system.

The career of Colonel George McDougall, Jr., the keeper of the Fort Gratiot Light at Port Huron, Michigan, on Lake Huron, from 1825 to 1842, perhaps best illustrates the potential for favoritism and corruption under this system of appointments. McDougall was the son of a prominent Detroit landowner. He became a lawyer in 1811 and served briefly with the militia in the War of 1812. He spent much of his life with bouts of ill-temper, excessive drunkenness, and the gout. He was constantly in trouble with the courts and was barred from practice several times. With his career and his health in a state of steep decline and his associates deserting him, McDougall convinced his friend William Woodbridge, the collector of customs at Detroit and later Michigan's Governor

and a United States Senator, to appoint him to the keeper's post at the newly opened lighthouse in Port Huron. The guaranteed lifetime salary of $350 appealed to McDougall, while his colleagues were delighted to see him exiled to a post some fifty miles from Detroit.

McDougall was a colorful keeper, to say the least. He stood five foot, nine inches tall, but weighed more than two hundred pounds and discovered during his first day on the job that he could climb the circular staircase leading up the tower only if he turned sideways. He could barely squeeze through the trap door leading up to the lantern floor, as it measured only eighteen by twenty-one inches. Once on the lantern floor, there was barely enough room for the lighting apparatus and this rotund keeper. He would never inspect the equipment again, but instead hired an assistant who would do all the work and thus, enable McDougall to have sufficient time to enjoy his vices. To help pay for his expensive style of living, he convinced Woodbridge to appoint him deputy collector of customs at an additional salary of $150 a year. He subsequently also held the job of postmaster in Port Huron, a position he nearly lost because he failed to submit quarterly reports to the Postmaster General. However, he was able to get another Michigan crony, Secretary of War Lewis Cass, to intervene on his behalf, so McDougall held three government positions up to his death in October, 1842. A contemporary recalled McDougall's deportment, hardly fitting for a public servant:

> *He took fits of drinking, and while they were on, no one could live with him. He was always accompanied by a colored youth, who was his valet, and seated comfortably in the bar-room of the hotel, his gouty foot resting easily on a cushioned chair, with his brandy toddy at his elbow, and his valet combing, oiling, and brushing out his voluminous wig, and cracking his jokes and making witty comments on the passing show, he was a picture to behold.*

Preceding page, although there is no positive identification of this handsome young lightkeeper available, he is believed to be Owen McCauley, who served for many years at the Squaw Island Light in northern Lake Michigan. Photo courtesy Owen McCauley family collection. Above, a photograph of the sitting room in the lightkeeper's dwelling at Eagle Bluff Light in Lake Michigan. This residence was restored to its original condition, and includes many of the original furnishings. At right, an interior view of the lightkeeper's house at the Old Presque Isle Light. This station is now a lighthouse museum.

Starting in the 1870s, the Lighthouse Board gained some control over its keepers. The Board accepted nominations from the collectors of customs, but did not make appointments until after a representative had interviewed the candidate. A three month probation followed before the keeper received a permanent appointment. The Board required that new keepers be between the ages of eighteen and fifty; be able to read, write, and keep simple accounts; be able to pull and sail a boat and perform other manual labor; and to have enough mechanical skills to maintain the equipment and do minor repairs. District inspectors, acting under the authority of the Lighthouse Board, could summarily dismiss keepers in two instances. Those found intoxicated were fired and ejected from the station, while any failure to keep the light burning was also grounds for dismissal, without regard to the circumstances or the keeper's previous service record. The keeper was expected to remain at his station as long as the

51

lighthouse was still standing.

The carrot was used far more often than the stick to produce good conduct. Although the laws governing the Lighthouse Service recognized only one grade of keeper, in practice, individuals were paid according to their length of service and the type of light station they operated. In the early 1890s, beginning assistant keepers might earn as little as $200 per year, while principal keepers responsible for the largest and most complex stations earned up to $1,000 annually. At minor pierhead or harbor lights, a single keeper was the rule. First and second order coastal lights had one keeper and one assistant, but if the station had a steam fog signal, a second assistant who was a steam engineer was added. At isolated offshore stations, as many as four assistants might serve. Recruits normally entered the Lighthouse Service as assistant keepers, and then were promoted or transferred to other stations as positions opened.

Scores of keepers had long, distinguished careers with the Lighthouse Service, often serving at a variety of light stations. However, they commonly remained on a single Great Lake over their entire career. Two Lake Michigan keepers are typical examples. Martin N. Knudsen received his first appointment as assistant keeper at Pilot Island after the previous assistant, John Boyce, committed suicide in June, 1880. Knudsen was appointed head keeper at South Manitou Island in 1881 and remained there until 1889, when he returned to

Pilot Island as keeper, a post he held until 1896, when the Lighthouse Service transferred him to a brand new station on Plum Island. He later ran the Racine Pierhead Light from 1899 until 1917, when he moved to the North Point Light near Milwaukee, the post he held when he retired in 1924 after forty-four years of service. Another Lake Michigan keeper, John Hahn, had a career which spanned the Lighthouse Bureau and Coast Guard eras. His first position was at Cana Island, off the Wisconsin coast, where he served a little more than a year starting in 1916. He then held posts at North Manitou Island for three years; worked eight years at the Sturgeon Bay Ship Canal Light on the Door Peninsula; spent ten years at the Manitowoc Pierhead Light; two-and-a-half years at the Holland Harbor Light; another two-and-a-half years at the Sturgeon Bay Canal Light; and finally, he ended his career working as keeper of the North Manitou Shoals Light Station from 1942 until January, 1950.

While the keeper's post was not hereditary, it was not uncommon for several generations of a family to serve as keepers. Philo Beers was the first keeper at the Grand Traverse, Michigan, light on Lake Michigan in 1853 and his son Henry took the post in 1858. Patrick Garraty became keeper at the Old Presque Isle Light on Lake Huron in 1861 and then served at the new light built nearby from 1870 to 1885. One son, Thomas Garraty, held the same post from 1885 until he retired in 1935. A

second son, Patrick, became assistant keeper at Presque Isle in 1890, but then served as keeper at the St. Clair Flats Range Lights from 1904 through 1917 and at the Middle Island Light on Lake Huron for twenty years, before retirement in 1937. His third son, John, worked as a keeper at several Lake Huron lights and finally, at the Mendota Light on Keweenaw Point, Lake Superior, when he retired in 1930. One daughter, Anna, was the keeper of the Presque Isle Range Lights from 1903 until her retirement in 1926. She is said to have sat in a rocking chair on her tiny front porch and rocked all night, watching the range lights to make sure they kept burning. Counting in-laws, the Marshall family had at least six keepers: Charles, at St. Helena Island light (1888-1900); his sons George at Mackinac Point (1890-1918), Walter at the Detroit lighthouse and son-in-law William Barnum, assistant keeper at Mackinac Point; George's sons, James, keeper at Mackinac Point (1918-1941) and Chester, keeper at Beaver Island and Manitowoc, Wisconsin.

Before the Coast Guard era, the majority of keepers were married men, but the Lighthouse Service did not specify that keepers be married. The general belief that a married man with a family would be more reliable and sober than a single person was nevertheless widely held. A young, single Irishman, John Malone, helped build the new Isle Royale (Menagerie Island) Lighthouse in 1875 and asked the district inspector for the post of assistant keeper. When the inspector indicated that he preferred a married man for the job, Malone promptly married and won the appointment. Grateful for the position, Malone then named his first child born at the light after the inspector. And in keeping with that tradition, Malone also named the rest of his twelve children, all born at the light, after the sitting district inspector. One year, when two inspectors held office, Mrs. Malone luckily had twins. But their string of good fortune ended when three inspectors served in a single year during the Spanish-American War. Usually, the

Above left, because of the hard to reach locations of many lighthouses on high cliffs, such as the Split Rock Light shown in this early twentieth century photograph, trams were often built to transport materials and supplies to the station. Photo courtesy Minnesota Historical Society. Above, not long after the lighthouse was built at Sand Point on Lake Superior, erosion of the shoreline forced the Lighthouse Board to move it inland. Here the work has just been completed. The station was pulled along a pathway made of large wooden planks. Photo courtesy Dr. and Mrs. Louis Guy collection.

keeper's wife and children shared in the work of maintaining the light and in a few cases were appointed assistant keepers. Mrs. Sarah E. Lane held the post of keeper at the Old Mission Point Light from the time of her husband's death in December, 1906 until July, 1908, when the Lighthouse Service appointed a male replacement. The practice of women holding positions was more common in the earlier days. According to one source, thirty women, mostly the widows or daughters of male keepers, served as keepers on the Great Lakes in 1851, when the Lighthouse Service operated a total of seventy-six lights on the Lakes.

Elizabeth Whitney Van Riper Williams was one of the most accomplished female lightkeepers on the Great Lakes. Her first husband, Mr. Van Riper, replaced Peter McKinley as keeper of the Beaver Island Light in 1869. McKinley was in chronic poor health and his daughters Effie and Mary had actually run the station for nearly nine years. In any event, Van Riper later died trying to rescue a ship's crew and his widow was appointed keeper a few weeks later. She married her second husband, Daniel Williams, in 1878, and she was subsequently appointed keeper at the new Little Traverse Lighthouse at Harbor Point, Michigan, in 1884. Elizabeth Van Riper Williams later published her autobiography, *Child of the Sea*, in 1905. In addition to Anna Garraty, previously mentioned, there was at least one other well-documented female lightkeeper, Miss

Harriet E. Colfax, who served at the Michigan City, Indiana, Light from 1853 to 1904, well into her eighties. Before the Lighthouse Board switched to kerosene as the principal illuminant, Miss Colfax heated lard oil on her kitchen stove during the winter months and hastily carried the hot oil up the lighthouse stairs to the lamp room, so she could light the lamp before the oil would begin to harden. One winter evening, during a violent storm, she had just descended the tower stairs and returned to the safety of her house when the wind blew the tower down.

The Lighthouse Service rules and regulations governing keepers were strict but not unduly rigid. Keepers could not take in boarders or pursue any trade or business which took them away from the station. As long as they performed their regular duties properly, keepers were allowed to engage in other work. The Lighthouse Board encouraged gardening and permitted work such as shoemaking, tailoring, and fishing. Keepers often served as preachers, justices of the peace, and school teachers. Blessed with a good supply of dead birds and a lot of spare time, one keeper at the Rock of Ages Light near Isle Royale became a highly accomplished taxidermist. The regulations covering absences from the station were so clearcut that a keeper could not easily hold down a job or run a business far from the station. Regulations clearly stated that keepers could leave the station only to draw their pay or to attend public worship on Sunday unless they had the permission of the district inspector. At stations with more than one keeper, only one could be absent at any time.

Beginning in 1852, the Lighthouse Board issued written rules of conduct for lightkeepers. These highly detailed instructions spelled out the required daily and monthly routines, and detailed procedures for cleaning, maintaining and repairing all of the equipment and apparatus. The Board's *Instructions and Directions for Light-House and Light-Vessel Keepers of the United States*, Third Edition (1858) was 87 pages long. It began with a

set of 25 general instructions for lighthouses with one keeper and 37 instructions for those with two or more keepers. Both sets began with the rules that the lamp should be lit at sunset, extinguished at sunrise and watched continually by the keeper throughout the night. The keeper also had to keep a daily log of the precise times the lamp remained lit, shipwrecks, the passage of ships in the area, the weather, and the consumption of supplies, particularly lamp oil. Regulation Thirty-Two outlined some general rules of conduct:

The light-keepers are required to be sober and industrious, and orderly in their families. They are expected to be polite to strangers, in showing the premises at such hours as do not interfere with the proper duties of their office, and may, with the approbation of the inspector, place a placard on a conspicuous part of the premises, specifying the hours when visitors will be admitted; it being expressly understood that visitors shall not be admitted to the lantern-room after sunset. No more than three persons shall have access to the lantern-room at one and the same time during the day; and no stranger visiting the light-house can be permitted to handle any part of the machinery or apparatus. The light-keepers must not, on any pretext, admit persons in a state of intoxication into the lighthouse.

Above left, Owen P. Young was the lightkeeper at Split Rock Light when this photograph was taken around 1925. Photo courtesy Minnesota Historical Society. Top, this photograph of Ed Lane, the lightkeeper for many years at the Michigan Island station in the Apostle Islands, was taken in 1939. Above, Bob Westfield also served at Michigan Island Light. Both photos courtesy Fran Platzke collection.

The *Instructions and Directions* included lists of general responsibilities, 11 for light-keepers and 34 for keepers of light-vessels. These included keeping the lamps, lantern glass, lantern room, and general premises clean and tidy. The written instructions became remarkably detailed when they dealt with the operation, cleaning, and maintenance of the lamps, lenses, and revolving mechanisms for the Fresnel lens apparatus. In a section running 34 pages long, the Lighthouse Board delineated 131 distinct procedures. The correct methods needed to install, trim and light the lamp wicks were spelled out, as

well as the expected height of the flame. Instruction XXI, "To Light A Burner With Concentric Wicks," leaves little to the keeper's imagination:

> *When the wicks are sufficiently saturated with oil, the lighting may be performed in observing the following precautions: The central wick, No. 1, is raised about 3/10 of an inch, and with the lighting lamp (article LXXV) four opposite points of the wick are lighted, and then lowered to the lowest point at which it will burn. Proceed in the same way with wicks Nos. 2, 3, etc., and hasten in each case to lower them as soon as they are lighted, so as not to smoke the apparatus. This being done, place the glass chimney on the burner, and put on the damper (article XI).*

A slightly less complicated set of 59 instructions also was included to take care of the surviving Lewis lenses equipped with Argand lamps and parabolic reflectors. The 1858 *Instructions and Directions* did not allow the keeper to use his own judgment, but must have eliminated mistakes and damage to the equipment. The detailed step-by-step procedures and the implied threat that the keeper would be dismissed if he departed from the instructions was perhaps the only way the Lighthouse Board could effectively run the far-flung system of aids with largely untrained personnel. As long as the keeper could read, was generally conscientious

and was willing to follow the instructions, the system would work reasonably well. Besides, quarterly visits by the district inspector served as another guarantee that procedures were correctly followed.

The inspector's visit produced much excitement and some anxiety for the keeper and his family. Martin Knudsen's children had detailed recollections of these visits in the 1890s. Once someone spotted the lighthouse tender carrying the inspector, the keeper and his wife had a few precious minutes to clean up and get ready:

> *Ma hastily looked over her little brood; a wash cloth here, a hairbrush there, and a clean blouse for that one, a fresh apron for herself. She and big sister then made a tour of the house to see that the beds were neatly made, extra shoes lined up in a row with the toes out, all clothes hung up as they should be, and a general order of neatness prevailing. The inspector looked into all of the corners. Pa and the assistants put the finishing touches on their chores and then donned their formal uniforms and caps.*

When the inspector arrived, Knudsen and his assistants stood at attention and saluted their superior, who then proceeded to examine the premises quickly, but thoroughly. Usually, the entire operation was done in a friendly, cordial, and totally professional manner. Before the visit ended, the lighthouse tender would leave a new library and remove the one left three months earlier. These inspections could not be predicted with much accuracy in the early days, so keepers could not afford to let major problems remain uncorrected very long. Once the telephone came into general use in the 1920s, keepers often phoned ahead to their comrades, giving several days' warning about an upcoming visit by the district inspector.

The Lighthouse Board made a conscientious effort to give the keepers a sense of pride in their work, a spirit of professionalism and the conviction that the Lighthouse Service would offer them a rewarding career. One move in that direction was the requirement, beginning on May 1, 1884, that all permanent male personnel wear a standard uniform. The Lighthouse Service supplied its keepers with their first set of uniforms, but then new and old employees alike had to bear the costs themselves. The new regulation stated,

> *The uniform for male keepers and assistant keepers of light-stations, and the masters, mates, engineers, and assitant engineers of light vessels and tenders, will consist of coat, vest, trousers, and a cap or helmet.*
>
> *The coat will be a double-breasted sack, with five large regulation buttons on each side—the top buttons placed close to the collar, the lower ones about 6 inches from the bottom, and the others at equal spaces between the top and lower buttons. It will be of the length of the extended arm and hand, and will be provided with two inside breast pockets and two outside hip pockets, the latter to have flaps so arranged as to be worn inside the pocket if desired. Each sleeve will have two small buttons on the cuff-seam.*
>
> *The vest will be single-breasted, without a collar, and cut so as to show about 6 inches of the shirt. It will have three pockets and five small regulation buttons.*
>
> *The cap will be made of dark blue cloth, with a cloth-covered visor and an adjustable chin-strap of cloth held by yellow-metal regulation buttons. A yellow-*

Above left, a photograph of the Emmanuel Lulich family. He was a keeper stationed at Sand Island Light in Lake Superior. Photo courtesy National Park Service. Above, Alphonse L. Gustafson was the lightkeeper at Devil's Island Light in the Apostle Islands from April, 1945 until he died on April 29, 1951. Photo courtesy Mrs. Lois Spangle collection.

Above, Captain E. J. Moore was the lightkeeper at Grosse Point Light from 1889 until 1922. Here, he is winding the clockwork mechanism, which rotates the light in the tower. Photo courtesy Evanston Historical Society. Above right, Mr. Balma oils the clockworks at Split Rock Light in Lake Superior. Photo courtesy Minnesota Historical Society.

metal light-house badge will be worn in the middle of the front of the cap. Masters of tenders will wear a gold-lace chin-strap instead of one of cloth.

Political influence on appointments ended only when keepers came under Federal civil service rules. The Pendleton Civil Service Act of 1883 gave President Chester A. Arthur authority to decide which Federal positions would be classified, requiring applicants to pass examinations. He did not put keepers into the classified civil service, but President Grover Cleveland added them through an executive order on May 6, 1896. From that time on, district engineers had to select new appointees from lists of eligible candidates maintained by local civil service boards. Shortly after it came into existence, the Bureau of Lighthouses in 1911 began granting keepers awards for the efficient performance of duties. Keepers who were considered efficient during four consecutive quarterly inspections in a single fiscal

year were awarded the "Inspector's Efficiency Star," which they could wear for a year. Keepers who won this award three years in a row were given the prestigious "Commissioner's Efficiency Star." These awards were a major factor in determining promotions, duty stations, and salaries. The Bureau also annually awarded the best-run light station in each district the "efficiency pennant," which could be flown at the station for a year. The awards created friendly competition among keepers and stations, contributing to an *esprit de corps*.

By the end of the 1920s, the number of Great Lakes lighthouses and keepers were already falling as automation and radio began replacing them. In 1925, the Tenth, Eleventh and Twelfth Districts, covering the Great Lakes, had a total of 778 permanent employees or thirteen percent of the total in the Lighthouse Service. There were 360 keepers and assistants on the Lakes, 230 light vessel personnel, 30 "light attendants," and nearly 100 construction and repair workers. All came under

Federal civil service regulations covering appointments, dismissal, and re-employment. Base pay for keepers and assistants varied according to length of service and the type of light station where the individual worked. Annual salaries of keepers ranged from $1,320 for an inexperienced man at a Class 1 Station (a minor light) to the top scale of $1,740 for a senior keeper at a Class 4 Station, one with a major light and a major fog signal. Assistant keepers' salaries similarly ranged from a low of $1,140 to a high of $1,560 paid to the senior assistant keeper at a Class 4 Station. Overall, the salary range was not very wide. An additional allowance of $60 a year was made for "moderate isolation" of the station, $120 for "extreme isolation," and $180 for "exceptional isolation, remoteness, and inaccessibility."

Keepers' fringe benefits improved considerably to the end of the Bureau of Lighthouses era in 1939. Through an Act of Congress in 1916, keepers and assistant keepers were entitled to free medical care from the U.S.

Public Health Service, a benefit enjoyed by light vessel personnel since 1886. The General Lighthouse Act of June 20, 1918, gave light keepers and lightship personnel their first retirement benefits. The Act permitted voluntary retirement at age 65, after 30 years service and compulsory retirement at age 70. The salaries of light vessel officers were increased substantially in 1924 to bring their pay up to the levels of comparable officers on private vessels. The pay schedule used by the Lake Carriers' Association was applied to Great Lakes light vessels. The Employees' Compensation Act of 1916 for Federal workers, plus the Lighthouse Disability Act of 1925, provided for the care of keepers disabled in the line of duty. After the Lighthouse Service became part of the Coast Guard in 1939, the civilian keepers slowly passed out of existence, disappearing by the mid-1960s.

THE KEEPER'S LIFE

In contrast to the romantic notions we often have about life at a lighthouse, many keeper's lives were isolated, lonely, routinized, boring, and occasionally dangerous. The keeper and family were often the pioneer inhabitants of a desolate coastal region or the only residents on a remote island. When Eber Ward was appointed keeper at Bois Blanc Island at the northern end of Lake Huron in 1829, he spent several weeks at the station alone before his family joined him. His nearest neighbors were on another island some eighteen miles away. During his term at this post, from 1829 through 1842, his wife left on only three occasions, twice to visit relatives in Ohio and once because she was seriously ill. Mail came once a month. He reported that, ''Indians often came to the island to fish, but were never troublesome, nor were we afraid of the other bands of forty or fifty drunken Indians that came sometimes.''

The Pilot Island Light (Port Des Morts Station), built in 1850 off the northern tip of the Door Peninsula on Green Bay, Lake Michigan, was on a rock often enveloped in fog for weeks at a time. The keeper at Pilot Island in 1872-1876, Lieutenant Victor Rohn, who had served in the Civil War, observed that the island offered ''as much independence and liberty as Libby Prison,'' but at least at the prison, one might have communication with the outside world. This desolate outpost depressed first assistant keeper John Boyce so much that he took his own life on June 20, 1880. And Ben Fagg, who visited the lighthouse in 1890, gave the following description:

> Pilot Island is a little island on 3½ acres of rocks and boulders on which there is an imported croquet ground, a few ornamental trees, a strawberry patch, two fog sirens, a lighthouse, a frame barn, a boathouse and some blue bell-shaped flowers and golden rods that grow out of the niches in the rock. It lies about two miles east of Plum Island, which is so called because it is plumb in the center of Death's Door.
>
> This is truly an isolated spot but I have spent five days on Pilot Island and they are among the happiest days of my eventuality. On moonlight nights it is like being in a dream of ideality to walk alone over the moss covered rocks and listen to the swish of the breakers that break over the breakwater at the boat landing, hear them roaring on all sides of the little island and to see huge vessels under full sails crossing the moon-glade on

their way through the Door. One seems to be completely separated from all that is worldly and bad. There is no field for gossip out here. The land is not suitable for general farming purposes, but it is a splendid place to raise an ample crop of good, pure thoughts.

One of the fog sirens at this station is an exact duplicate of the one that was on exhibition at the World's Fair held in Paris. Its song can be heard a distance of forty miles, and when it sings all the lights in the signal house must be hung by strings to prevent them from going out. The sound is so intense that no chickens can be hatched on the island, as the vibration kills them in the egg, and it causes milk to curdle in a few minutes. Visitors at the lighthouse on foggy nights sit up in bed when the siren begins its lay and look around for their resurrection robes.

Lake Superior lighthouses were more isolated than those on the other lakes because of the combination of greater distances and more severe climate. Charles Davis, the keeper of the Copper Harbor, Michigan, range lights, at the northern end of the Keweenaw Peninsula, noted in his diary on November 28, 1891, the terrible conditions experienced by his fellow light keeper at Gull Rock, off the tip of the same finger of land: "Gull Rock Keeper Nolan is still at the Rock, his wife too. It is bad enough for men to do any boating this season with such weather as we are having, but the idea of keeping a woman at our isolated station this late, does not show very good judgement, when the thoughts of an open boat on Lake Superior will make the hardiest of men shiver." In November, 1883, the keeper at the Passage Island Light, at the northeastern tip of Isle Royale, went to Port Arthur for supplies, but was not able to return as winter set in. He left his Chippewa Indian wife and their three children on the island and they spent the entire winter there alone. She fished and snared rabbits for food, but when her husband finally returned in the spring, they were barely alive.

Above left, A. G. Carpenter was stationed at Outer Island Light from 1935 until he was transferred to the station at Raspberry Island in 1940. He then served as keeper of that light for four years. Photo courtesy Fran Platzke collection. Above, a photograph of the keeper and his family stationed at Sand Island not long after it was built in 1881. Photo courtesy National Park Service.

Stories of isolated keepers at Lake Superior stations are plentiful. The keeper of the Outer Island light in the Apostle Islands died during the last few weeks of one shipping season. The assistant keeper was unable to contact passing vessels and spent two weeks with the corpse until the lighthouse tender finally came and took them off. The isolation of many lights produced many interesting incidents, not always disastrous. James Corgan, keeper of the Manitou Island light on Lake Superior, not far from Gull Rock, noted in his logbook on July 15, 1885: "Principle keeper started at 8:00 p.m. in the station boat with wife for Copper Harbor (distant 14 miles), in anticipation of an increase soon after arriving. When one-and-one-half miles east of Horseshoe Harbor, Mrs. Corgan gave birth to a rollicking boy; all things lovely, had everything comfortable aboard. Sea a dead calm."

The journal of keeper John Malone, who served at the Isle Royale Lighthouse between 1878 and 1893, illustrates the life of a keeper at an isolated Lake Superior light station. He normally left the island for the mainland (Houghton, Michigan) between November 20 and December 1, returning to relight the station the first week in May. Malone's journal was filled with storm reports, such as the one for October 16, 1880: "Hail, snow and rain—a tempest; Lost boat, boat house wayes, and dock. It was impossible to save anything. I don't believe a cat could get from the dwelling to the boat house. It hailed awful." In the spring and summer of 1884, the weather was so severe that Malone's family stayed on the mainland until late July. On October 28, he reported that the island "looked like an iceberg." By November 10, he began sounding desperate: "It is almost impossible for us to stay here much longer for we have to cut the ice from the way every day or we could not launch our boat, the only hope we have of getting off for winter quarters." They got off the island on November 16 and went to Duluth for the winter. The next spring, Malone made the following entry on May 11, 1885: "Arrived here at 12 o'clock and commenced lighting up. We had to keep three stoves going night and day steady to heat up the house. It was just like an ice house. The lake is full of ice and the weather is very cold." And finally, at the beginning of the following winter, Malone's comment on November 4, 1885, told all—"This must be the North Pole."

Keeper Malone had time to watch the birds, noting that "A great flock of white owls passed by the island" on June 12, 1884 and that "the geese are making for winter quarters" on October 5, 1885. He and his assistant usually found plenty of fish and game to eat. On July 31, 1884, he caught four lake trout, the largest weighing 19 pounds and the smallest 12 pounds. His assistant, John, caught 36 brook trout on Isle Royale on July 13, 1886 and another 31 in early October. His assistant also shot three black ducks and four rabbits on November 2, 1886. In June of 1890, he and his assistant caught five hundred pounds of lake trout and "got three cents per pound." On August 31, 1890, "the assistant picked one-and-a-half pails of Rosberries and killed two prairie chickens." One of their most important sources of food each spring were the gulls. Malone reported on May 13, 1886: "Sea gulls commenced laying eggs. We got 12 on Menagerie Island. We got 357 eggs off the rocks at Siscowitt Point." They collected more than a thousand eggs in May, 1886 and then ate about thirty a day in June and early July. The following spring they

had collected a grand total of 1,478 gull eggs by the first of June. Malone provided detailed counts of their gathering of gull eggs every spring. As of May 18, 1892, he reported: "We ain't found any gull eggs yet, this spring." By the end of May, however, they had collected 1,250 eggs.

The capricious and often violent Great Lakes weather made the keeper's life difficult and dangerous. Many stations had ice problems, especially at the close of navigation, but the Stannard Rock Light, some twenty-three miles southeast of Manitou Island in Lake Superior and known generally as the "loneliest place in America," had the biggest ice jams of any light. When the lighthouse tender arrived at Stannard Rock to take the men off at the end of the season, it was often so iced up that the keepers had to walk several hundred feet across the ice, climb into a small open boat and then be transported to the tender, anchored out of danger. After a major ice storm there in late 1913, the Stannard Rock

Above left, Hans Frederick Christensen was the lighthouse keeper in charge at the Superior Entry (Minnesota Point) Light from April, 1934 until November, 1939. Soon after, this light station was abandoned by the Coast Guard. Photo courtesy Elmer R. Christensen collection. Above, an unknown keeper lights the lamp at the Grosse Point Light on Lake Michigan. Photo courtesy Evanston Historical Society.

pier became a giant ice sculpture, with the base of the light tower covered with a layer of ice twelve feet thick. That year, a rescue party of twelve men worked from the eleventh of November through the fifteenth to free the trapped keepers, using steam lines, shovels, and axes to break through.

The worst weather ever experienced at Stannard Rock involved storms which began on November 22, 1929 and lasted eleven days. Besides snow squalls, the major problem was icing caused by the combination of gale-force winds and temperatures of ten below zero. The tower was encased in ice from the top of the lantern down to the water and the winds were so strong that the keeper had to replace the oil vapor lamp with an old wick lamp which he tied down with wire. He used a steam hose to keep the fog signal trumpets thawed enough to operate if needed. The tender *Amaranth* took the keepers off Stannard Rock on the fourth of December. Ice problems were by no means limited to Lake Superior. In a storm in January of 1928, the Ashtabula Harbor Light on Lake Erie was encased in a solid mass of ice five feet thick, trapping the keepers inside for several days.

Getting away from the light station at the end of the season was often attempted under dangerous conditions. Many keepers died after leaving their stations in small boats and running into the violent late fall storms found on the Great Lakes. At the end of the shipping season of 1900, five died this way on Lake Michigan alone and the Lighthouse Service advised keepers to wait for a steamboat or tug to take them off. On December 14, 1900, five people living at the Squaw Island Light in northern Lake Michigan left for St. James on Beaver Island, nine miles away, in a twenty-five foot sailboat. The unfortunates were Captain William H. Shields, the keeper, and his wife; the two assistant keepers, Lucien F. Morden and Owen C. McCauley; and Mrs. Lucy Davis, a niece of Shields. They ran into a sudden squall, which capsized the boat, throwing everyone into the water. Morden and both women died in short order, but the other men, lashed to the wreckage, drifted for twenty-three hours, covered with ice, before the crew of the steamer *Manhattan* rescued them. The two lived and the Lighthouse Service later appointed Shields, who lost a leg in this tragedy, keeper of the Charlevoix, Michigan, lighthouse supply depot. Owen McCauley was appointed chief keeper at Squaw Island.

Lake Superior naturally presented the greatest dangers at the end of the season. At the beginning of the winter of 1919, the lighthouse tender *Marigold*, captained by J.N. Lanstra, made several daring and dangerous rescues of keepers at Lake Superior light stations. They took off the keepers at Stannard Rock on the second day of December, then waited out severe weather before relieving the keepers at Raspberry Light in the Apostle Islands a week later, just as they were about to run out of food and coal. The ship continued its mission and

rescued the keepers at Sand Island and Outer Island, both in the Apostles, on December 11. They reached the Rock of Ages Light on December 16, by which time the food supply there was down to a single can of tomatoes. At Rock of Ages, Lanstra reported, "two of the assistant keepers were frost bitten while coming out to the tender, but our hard boiled seamen stood the hardship all right." When the *Marigold* returned to Duluth, the ship was nearly out of coal and faced a solid mass of ice extending two miles out from the city. Lanstra and his crew rammed their way into port and were ready to go out again to retrieve the last remaining keepers, but found that they were safe. A Canadian icebreaker had rescued the keepers off Passage Island, while the men on Devil's Island in the Apostles had managed to walk ashore over Lake Superior's frozen skin.

The Lighthouse Board was keenly aware of the morale problems caused by the extreme isolation and loneliness of some stations. Starting in 1876, the Board assembled small, portable libraries containing about fifty volumes, enclosed in a case. Initially, the library was left at a station for six months and then replaced by a new one from another station. These "lighthouse libraries" became so popular that the exchange took place every three months when the district inspector made his quarterly visit to the station. In the first year of this program, the Board collected fifty libraries, but by 1888, the number in use was about 550 altogether. The Bureau of Lighthouses still had 350 libraries in use as late as 1912 and many survived into the 1920s. The individual library typically included a mixture of history, fiction, poetry, scientific works, and a Bible.

The education of the keeper's children was a major problem at many stations. The Lighthouse Service generally gave preference to keepers with school-age children when filling vacancies at stations convenient to schools, while trying not to place those families at stations totally isolated from schools. Keepers sometimes had to board their children at great distance and expense

Above left, a photograph of a gathering of lightkeepers' wives and families who were stationed in the Apostle Islands in Lake Superior. Photo courtesy National Park Service. Top, a group portrait of the keeper and assistant keepers, and their wives and children at the Raspberry Island Light. Photo courtesy National Park Service. Above, lightkeeper Earl Seseman with his wife Thrya, pose with first assistant Eino Hill and his wife Almi. They were stationed at Raspberry Island in the early 1940s. Photo courtesy Fran Platzke collection.

to ensure their education. In the state of Maine, traveling teachers helped educate lighthouse keepers' children. In some cases, the parents served as teachers, but on one occasion, such efforts resulted in tragic consequences. The children of the keeper at St. Martin Island on Green Bay rowed over to Washington Island every day to attend a school there. One day they disappeared without a trace and since that day, on dark, windy nights, an eerie green light is often seen moving slowly along the southwest shore of St. Martin Island, supposedly the ghost of the lightkeeper, carrying a green lantern, still searching for his lost children.

Another well-documented lightkeeper's ghost lived at the Waugoshance Light just west of the Straits of Mackinac. John Herman, the keeper there from about 1885 through August, 1894, drank heavily while on shore leave and his condition often persisted for several days after he returned to duty. His assistant lit the lamp one evening at sunset and found himself locked in the lamp room, a victim of one of Johnnie Herman's practical jokes. The imprisoned man called down to Herman, also well-lit, and staggering along a pier, but the jokester disappeared, perhaps deciding to cavort with the fish. According to local gossip, from that day on, strange things began to happen at Waugoshance. If a keeper fell asleep on duty, his chair was kicked out from under him. Doors would mysteriously open, or worse, get closed and locked. Some unseen being

reportedly shoveled coal into the boilers. And all of these incidents were presumably the handiwork of the mischievous ghost of Johnnie Herman. After the White Shoal Light went into service in 1910, the Lighthouse Service closed Waugoshance, claiming it was obsolete. But everybody familiar with events there knew they had to close the light at Waugoshance because nobody wanted to contend with the ghost that inhabited the place. The man who had loved spirits most of his life became one after death.

A lightkeeper's lust produced at least one ghost, a wailing woman who inhabits the ''new'' Presque Isle Light on Lake Huron. Her husband is said to have found a girlfriend in a nearby town and when he visited her, he would lock his wife in the tower for safekeeping. In time, the story goes, the keeper murdered her to end her complaining. He explained her sudden absence from the station by claiming that she had returned to her family in the South. But according to the legend, his victim chose to remain in the tower, at least in spirit, to torment her husband and others who have lived there with her long spells of crying and moaning.

HEROINES AND HEROES

Lighthouse keepers and their families often showed courage and fortitude in protecting and operating the light station. Emily Ward, the daughter of Eber Ward,

keeper at Bois Blanc Island on Lake Huron, was marooned on the island practically by herself during a violent winter storm in January 1838. A young boy named Bolivar, whom the Wards had adopted a few years earlier, was there as a companion, but spent most of the time quivering with fear. As the storm worsened, high waves began to pound at the base of the tower, causing the brick walls to crack and buckle. Realizing that the structure would not last much longer, Emily made five trips up the 150 stairs to the lamp room and down again, carrying the lamps, reflectors, and lenses to a safe spot well away from the tower. She and Bolivar then abandoned the keeper's dwelling, which was attached to the tottering tower, and from the woods a safe distance away, they watched the tower fall over. In a later incident, James Davenport, one of the keepers at Waugoshance, was stranded alone at the light during the time of the Great Chicago Fire of 1871. The wind from the southwest carried thick smoke all the way up to the

Above left, late winter ice still surrounds the light at Racine Reef, Wisconsin, as the keeper and crew attempt to put a boat in the water. Photo courtesy United States Coast Guard. Above, a photograph of the work crew washing clothes at the bottom of the tram below the construction site of the Split Rock Light station. Photo courtesy Minnesota Historical Society.

Above, a photograph of the keepers and assistant keepers from the five light stations located in the Apostle Islands in western Lake Superior. Photo courtesy National Park Service. Top, a portrait of Harriet E. Colfax who served as the lightkeeper of the Michigan City, Indiana, Light from 1861 until 1904. Photo courtesy Michigan City Historical Society. Above, a photograph of Elizabeth Whitney Williams taken from her autobiography Child of the Sea. *She served as lightkeeper at Beaver Island, as well as at the Harbor Point Light.*

Straits of Mackinac, where it lingered for three days, worse than any ordinary fog. The station had a rope-operated fog bell and Davenport sounded the bell for three days without rest. He sat in a chair with the rope in one hand and several pots and pans on his lap, so that when he dozed off, the pans hit the floor and woke him up. Because of his efforts, only seven ships ran aground on the nearby reefs during this long ordeal.

Keepers almost routinely rescued mariners in nearby waters. On May 11, 1890, a man capsized his rowboat on the Detroit River near the Mamajuda Lighthouse. The keeper, Orlo Mason, was away, but his fourteen-year-old daughter, Maebelle L. Mason, launched the family boat and rowed more than a mile to the drowning man, hauled him into her boat and rowed back to the lighthouse. She received the Silver Lifesaving Medal for her courage and later, the Ship Masters' Association presented her with a Gold Medal and Maltese Cross. Two years later, on November 9, 1892, Martin

Knudsen, the keeper of the Pilot Island Light on Green Bay, rescued six people from the *A.P. Nichols*, wrecked on the rocks off Pilot Island. Knudsen walked out to the ship in the pitch black night by way of several shoals, with the raging seas up to his neck, managed to get the five men and one woman off the boat and then guided them safely back to the shore. Another schooner, the *J.C. Gilmore*, had run aground on the island a week earlier, and keeper Knudsen had to feed sixteen lighthouse guests for more than a week before the weather finally improved and they left the island.

Lake Superior had more than its share of shipwrecks and rescues. On December 5, 1906, the Canadian passenger ship *Monarch* was driven aground off Isle Royale by a violent snowstorm. The forty-one people aboard managed to get ashore, but had no food or shelter and suffered badly. They were able to start a fire and luckily, on their third day marooned on the island, a bag of flour washed ashore and they made some bread. Finally, the assistant keeper of the Passage Island Light, Klass Hamringa, saw the smoke and rowed his small boat seven miles to the scene of the wreck. He could not get ashore because of the high seas, but Purser Beaumont of the *Monarch* swam out to the rowboat and the two men went back to the lighthouse. Once there, they contacted the tug *Whalen*, which rescued the survivors after they had spent a total of four days on Isle Royale. Many had suffered from frostbite and one man died. In November 1914, the keeper at the Whitefish Point Light and two fishermen joined forces to rescue eleven men from the *Ora Endress*, which had capsized near Whitefish Point. Dozens and dozens of less dramatic rescues were common, but rarely received much attention. When called to action, lightkeepers showed the same courage and fortitude as the men of the Lifesaving Service. A lightkeeper or a member of his family rescued someone on the Great Lakes nearly every year and they saved thousands of lives simply keeping their lights and fog signals in service. □

5 LAKE ERIE AND LAKE HURON LIGHTS

The discovery of the Great Lakes did not take place in the order one might expect. Lake Erie was the last of the Great Lakes to be seen by white men, while Lake Huron was the first. Samuel de Champlain, founder of Quebec, reached Georgian Bay via the French River in July, 1615 and named the bay "the Freshwater Sea." By 1640, the French explored the upper parts of Lake Huron and had ventured into Lake Michigan and Lake Superior. Finally, in 1669, Louis Jolliet descended Lake Huron, passed through the St. Clair and Detroit rivers and reached Lake Erie. The French soon recognized the strategic advantage of controlling the Detroit River and founded Detroit as a military outpost in 1701.

With the rapid growth of commerce serving Lake Michigan from the 1830s on, and the opening of the Soo Locks in 1855, the narrow links between Lakes Erie and Huron, consisting of the Detroit River, Lake St. Clair, and the St. Clair River, became more congested. Detroit flourished as a result, serving as a convenient supply point for cordwood, provisions, and sailors.

As Detroit and other cities along the Great Lakes shoreline expanded in the late nineteenth century, residents began to also use the Lakes and connecting rivers for leisure. Pleasure cruising, mainly on private excursion boats, became enormously popular beginning in the 1890s and this form of recreation continued to draw large crowds through the 1940s.

At the height of their popularity, hundreds of large excursion boats plied the Great Lakes in the early decades of this century. The *Tashmoo* (1900-1936), a White Star Line vessel based in Detroit, normally ran a round trip each day from Detroit to Port Huron, making twenty stops enroute, and having an overall running time of five-and-one-half hours one way. Other popular vessels such as the *Put-In-Bay* (1911-1953) sailed from Detroit to Put-in-Bay on Bass Island in western Lake Erie, then on to Sandusky, Ohio and back. This cruise featured a fine evening of dancing to the popular music

of the day in an enormous ballroom. The Detroit and Cleveland Navigation Company operated the *City of Detroit III* (1912-1956) between Detroit and Buffalo for most of its history, while other vessels made regular trips to Mackinac Island, Lake Superior, and Chicago. But not all of these pleasure boats are gone. The two ships built to ferry passengers to Bob-Lo Island, the *Columbia* (1902) and the *Ste. Clair* (1911), remain in service and continue to bring pleasure to the thousands of visitors to Bob-Lo Island each summer.

West Sister Island Light (1848, 1868)

Originally built in 1848, this conical stone tower was renovated and raised in 1868, when a new keeper's dwelling connected to the tower by a covered passage, was also built. The light station originally displayed a Fourth Order Fresnel lens, but it now has a 300 millimeter plastic lens encasing a battery-powered light. The last keeper left the island in 1937 when the station was automated. The keeper's dwelling was destroyed during the Second World War when the U.S. Army used the island for artillery target practice. All the other original buildings are gone as well, and the island now serves as a national wildlife refuge.

Manhattan Range Lights (1918)

Range lights are used by mariners to fix their position in open water, and to help them guide vessels into port. These lights typically appear in pairs, normally at least 1,000 feet apart, and the focal plane height of the rear light is always much higher than that of the front light. The Manhattan Range Lights, which guide ships into the Maumee River, were constructed in 1918 by the Great Lakes Dredge and Dock Company. They replaced an older set of range lights built in 1895. Both are located atop skeletal steel towers 1,300 yards apart. The front light has a focal plane 40 feet above the water level,

POINT AUX BARQUES LIGHT

WEST SISTER ISLAND LIGHT

72

but the rear light is some 86 feet above the river. The rear light tower houses a Fifth Order bull's-eye lens fitted with a spherical brass reflector, while a piece of red plastic suspended over the lantern gives the light its characteristic color. A keeper was once stationed in a nearby dwelling, but this light is now automated.

Toledo Harbor Light (1904)

After the shipping channel into Toledo Harbor was enlarged in 1897, traffic increased substantially, and a permanent light to mark the channel was needed. Construction of this light and fog signal station began in 1901, and the light was first exhibited on May 23, 1904. The lighthouse structure is on a concrete base, which rests on a submarine crib filled with stones. The Romanesque-style, buff-colored, three-story brick building, as well as the attached one-story fog signal building, have steel frames. The main building had apartments for the keeper and his two assistants. The cylindrical lantern encasing the light measures 8 feet, 6 inches in diameter and rests atop a cylindrical tower some 13 feet in diameter. The Third-and-a-half Order lens, made by Barbier & Benard of Paris, consists of a large bull's-eye of 180 degrees, with a half cylinder of ruby glass, and two smaller bull's-eyes of 60 degrees each. When rotated, the lens produces two white flashes followed by a single red flash. The apparatus originally

was turned by a clockwork mechanism driven by weights, but when it was automated by the Coast Guard in 1965, an electric motor was installed.

Detroit River Light (1885)

Beginning in 1837, the Lighthouse Board urged the construction of a light on or near Bar Point Shoal, located in Lake Erie just south of the entrance into the Detroit River at a point where ships made the turn to head into the river. Many vessels had run aground there and the Canadian Government finally established a lightship on the Bar Shoal in 1875. Unfortunately, the accidents continued because the lightship was not very powerful and the configuration of lights in this area often confused navigators, for there were lights on both the American and Canadian shores in addition to those aboard the ships passing through this narrow stretch of water. Congress finally appropriated a total of $60,000

for this project in 1882 and 1883, but an additional sum of $18,000 was needed before the work was completed in 1885. The entire project was designed and supervised by Captain C.E.L.B. Davis of the U.S. Army Corps of Engineers. Although test borings were made at the site in 1882, actual construction began in 1884. A timber crib measuring 45 feet by 18 feet high was built at Amherstburg, Ontario, and towed to the site, where it was sunk in 22 feet of water on July 3, 1884. The crib was then filled with concrete, a task that took more than two months. Next, they constructed the pier consisting of cut stone blocks and measuring 43 feet by 90 feet by 15 feet high, with 4 feet below the water line. The pier was completed in late November, but at that point the settlement of the crib and pier had been uneven, some 16 inches out of level. Before stopping work for the winter, the work crew loaded 550 tons of rubble stone onto the pier, mostly on the high side and when they returned the following spring, the pier was level. They

DETROIT RIVER LIGHT

GROSSE ILE NORTH CHANNEL FRONT RANGE LIGHT

WILLIAM LIVINGSTONE MEMORIAL LIGHT

WINDMILL POINT LIGHT

proceeded to build the light tower and fog signal building, and the light was first exhibited on August 20, 1885. The conical tower is built of cast iron plate, surmounted by a round watchroom and a 10-sided cast iron lantern with an inscribed diameter of 7 feet, 4 inches. The tower is 49 feet high, 22 feet in diameter at the base, and 18 feet in diameter at the parapet. It is basically the same as the tower erected at Harbor Beach in 1885. This station also includes a fog signal building. The lens exhibited here has been changed several times in order to change its characteristics. In 1907, for example, a lens with a single fixed panel of 180 degrees and six bull's-eye panels of 30 degrees each was installed, so that the light was fixed for 30 seconds and then produced six flashes at five second intervals, with the entire lens rotating once a minute. The lens now exhibited has six panels of 60 degrees each, with three bull's-eye flash panels each separated from the other by a 60 degree blind panel.

Grosse Ile North Channel Front Range Light (1906)

The original lighthouse at this site was built in 1894 and stood on three wooden stilts or legs. The surviving tower is an octagonal wooden structure standing forty feet in height and resting on a concrete foundation. The Coast Guard decommissioned the light in 1963 and sold the property to the Grosse Ile Historical Society, the current owners.

Detroit Lighthouse Depot (1874)

Beginning in 1864, the Lighthouse Board built a central supply depot on Staten Island, New York, to store illuminating oils and equipment, buoys, lenses, fuels, and other supplies. The Board eventually established similar depots in each of the twelve lighthouse districts, with Detroit receiving the first depot on the Great Lakes. In 1869, the United States Marine

Hospital gave part of its property on Mt. Elliott Avenue to the Treasury Department for the depot. The Lighthouse Service began building a permanent warehouse there in 1871, but it was not finished until 1874. The resulting storage building, still standing and largely unaltered, is an impressive three-story brick building measuring 40 by 60 feet, with a gabled roof supported by iron trusses. This handsome structure was part of the U.S. Coast Guard Group Detroit headquarters until recently, when the Federal Government transferred it to the City of Detroit, which plans to use it as a museum.

William Livingstone Memorial Light (1929)

This monumental white marble tower stands about 70 feet tall and exhibits an unusual occulting white light which is visible for 16 miles to the east. It was built in 1929 with private donations as a memorial to William Livingstone who was President of the Lake Carriers' Association from 1902 until his death in 1925. Livingstone promoted many improvements in the navigation of the Lakes and argued for the construction of a separate down-bound channel on the lower Detroit River. And when it was finally built, the Federal Government named the channel after him.

Windmill Point Light (1933)

The first permanent Federal lighthouse was built here, at the head of the Detroit River, in 1838, after numerous complaints from mariners about the difficulty of navigating this narrow passage at night. The light tower was substantially rebuilt in 1866, 1875, and again in 1891, but the tower standing today was erected in 1933. The white conical structure encased in steel plates holds a flashing white light with a focal plane 42 feet above the lake level. The lantern is fitted with a Sixth Order lens, the smallest size Fresnel lens that was made.

LAKE ST. CLAIR LIGHT

FORMER PECHE ISLAND REAR RANGE LIGHT, NOW AT MARINE CITY

ST. CLAIR FLATS OLD CHANNEL FRONT RANGE LIGHT

ST. CLAIR FLATS OLD CHANNEL REAR RANGE LIGHT

Lake St. Clair Light (1941)

This modern structure is a white, round, steel tower covered with steel plates resting on an octagonal concrete base. The tower exhibits a flashing white lens lantern which is 53 feet above the level of the lake.

St. Clair Flats Old Channel Range Lights (1875)

This pair of range lights was established to guide vessels into the channel leading from Lake St. Clair into the St. Clair River. Congress first authorized funds for this work in March, 1853, but the two lights were not completed until 1859. The original rear range light, a conical brick tower resting on a stone pier, has survived intact. It stands about 40 feet tall and has a 10-sided cast iron lantern. The front range light tower, however, started to lean badly as its foundation began to settle, and in 1875, the Lighthouse Board built an entirely new

crib, tower, and keeper's dwelling for $10,000. The conical yellow brick tower, measuring 11 feet in diameter at the base, is 17 feet tall and rests on a submerged timber crib 30 feet square and 6 feet tall. Virtually all of the 10-sided cast iron lantern is gone and the structure is now fitted with a plastic lens. History seems to be repeating itself, for once again, this tower is leaning noticeably.

Peche Island Rear Range Light (1908)

This light was originally erected in 1908 as the Rear Range Light at Peche Island in the Detroit River near Belle Isle. It is a round tower of steel plate construction, topped off with an octagonal lantern having a Sixth Order Fresnel lens. Because the tower had been leaning for several years, the Coast Guard had it removed in October, 1982 and replaced it with a new tower. The old light is now part of a maritime park in Marine City.

Lightship Huron

View along the deck of the Lightship Huron

Lightship Huron (1921)

This lightship, originally commissioned in 1921 as Light Vessel No. 103, was built by the Charles L. Seabury Company of Morris Heights, New York, at a cost of $147,428. It went into service as a relief vessel used to replace other lightships on duty in northern Lake Michigan. This ship became the Gray's Reef Light Vessel in 1923-27 and 1930, and served at North Manitou Shoal for the 1934 season. The following year, she became the permanent Huron Station Light Vessel, assigned to the Corsica Shoals, six miles north of Port Huron. She remained on duty there until 1970, and for the last thirty years of her commission, the *Huron* was the only American lightship on the Great Lakes. Overall, she is 97 feet long, with a beam of 24 feet, and a displacement of 340 tons. In 1949, her original steamdriven engines were replaced by diesel engines. After decommissioning the *Huron* in 1971, the Coast Guard gave the vessel to the City of Port Huron, and the lightship, now stationed on shore, is open to visitors.

Fort Gratiot Light (1829, 1861, 1875)

This light was the first lighthouse on Lake Huron, built in 1825. It was constructed just north of Fort Gratiot, an outpost established in 1814 to protect the American border against British attack. Destroyed by a violent storm in September, 1828, this station was rebuilt by Lucius Lyon at a cost of $4,445, and the light was put back in service in December, 1829. The tower was subsequently raised 20 feet in 1861 to its present height of 86 feet. The adjacent two-story lightkeeper's dwelling was built in 1874-75 to house two families. The first keeper, Rufus Hatch, served only six months until he died. His replacement, Colonel George McDougall, held the post from 1825 until 1842. This lighthouse was automated by the Lighthouse Service in 1933, but it has

FORT GRATIOT LIGHT

PORT SANILAC LIGHT

retained most of its original architecture and equipment, and is the oldest surviving lighthouse in Michigan.

Port Sanilac Light (1886)

Congress appropriated $20,000 for the construction of a coastal light at Port Sanilac in 1884, and work began in April, 1886. The light was first exhibited on October 20, 1886. The station includes an attached brick keeper's dwelling, which is now privately owned. The octagonal brick tower stands 59 feet tall from the base to the top of the ventilator ball and creates a lens focal plane 69 feet above the mean low water level of Lake Huron. The tower has a diameter of 14 feet at the base, but then tapers to a diameter of 9 feet at the parapet. The octagonal cast iron lantern is fitted with a Fourth Order Fresnel lens manufactured by Barbier & Fenestre of Paris, consisting of nine fixed panels and a brass reflector, creating an arc of illumination of 300 degrees.

Harbor Beach (Sand Beach) Light (1885)

In 1885, the U.S. Army Corps of Engineers finished building a harbor of refuge at Sand Beach, a long-needed facility because there was no harbor on Lake Huron between Port Huron and Saginaw Bay. Begun in 1873, the harbor project cost nearly $1 million. This light tower was built on a timber crib measuring 38 feet by 57 feet by 28 feet high, with the tower virtually identical to the Detroit River Light completed the same year. It is a conical brick structure encased in cast iron plates, measuring 22 feet in diameter at the base and 18 feet in diameter at the parapet. The brick walls are two feet thick at the base. The tower is surmounted by a round cast iron watchroom, which supports a 10-sided cast iron lantern, creating a lens focal plane 54 feet above the mean low water level of Lake Huron. The lens, which bears the inscription, "Barbier & Fenestre, Constructeurs, Paris, 1884," consists of ten panels, each

POINT AUX BARQUES LIGHT

with a bull's-eye flash panel in the center. This Fourth Order Fresnel lens is now turned by an electric motor.

Pointe Aux Barques Light (1857, 1908)

Congress appropriated a sum of $5,000 in 1847 for the construction of a lighthouse at Pointe Aux Barques ("point of the little boats") to mark the turning point from Lake Huron into Saginaw Bay, and to warn mariners of the shallow waters along these coasts. The original lighthouse was a stone building, but it was rebuilt in the mid-1850s by the contractors Alanson Sweet, Luzene Ransom, and Morgan Shinn, who won contracts from the Lighthouse Board to rebuild several light stations in Michigan at the same time. They completed their work in 1857, and the assistant keeper's house was later added in 1908. A lifesaving station established nearby in 1876 now serves as a museum, but this station was moved from its original location. The light tower is 89 feet tall and creates a lens focal plane 93 feet above the water level. The flashing white light can be seen for eighteen miles and it remains in service, although the station is now an automated, unmanned facility. The Coast Guard sold the land and the keeper's house to Huron County in 1958 and the county maintains the property as a park. Dan McDonald, the last civilian keeper at Pointe Aux Barques, described the work of a keeper just prior to his retirement in 1955:

PORT AUSTIN REEF LIGHT

"We sat through lonely vigils in the tower of the lighthouse, keeping a constant lookout for changing weather conditions and ships which might come too close. We were ready for anything—a ship which might run aground, a freighter needing to bring a sick seaman ashore, or a swamped fishing boat which needed help."

Port Austin Reef Light (1878, 1899)

Port Austin Reef extends into Lake Huron at the tip of Michigan's ''Thumb,'' the turning point for northbound ships heading into Saginaw Bay. Congress appropriated $10,000 in March, 1873 and approved an additional $75,000 in March, 1877 for the construction of a light to mark this reef. A total of $81,871 was finally spent on the project and the light was first exhibited on September 15, 1878. The light station originally rested on an octagonal pier 80 feet in diameter, 33 feet on each side, and 29 feet high, but had only six feet under the

water line. The shape of the pier was modified in 1899 with the addition of a new section attached to the existing pier. At the same time, the station was rebuilt to its present configuration. The fog signal building, a brick structure measuring 34 feet square, with a gabled roof, is attached to the yellow brick light tower, which is 16 feet square and 60 feet tall from the base to the top of the ventilator ball. Because of the height of the pier, the lens focal plane is 76 feet above the mean low water level of Lake Huron. The tower has an inner wall 4 inches thick, surrounded by an air space of 3 inches, and an exterior wall 13 inches thick. It is surmounted by a round cast iron lantern, 7 feet, 8 inches in diameter, with helical bars across the glass lantern panels. After the 1899 reconstruction, this lighthouse exhibited a rotating Fourth Order Fresnel lens consisting of five flash panels and two fixed panels, manufactured by Henri Le Paute of Paris. The lantern is now fitted with a 200 millimeter glass lens. The station did not provide

Saginaw River Rear Range Light

a permanent dwelling for the lightkeeper, because he normally lived on the mainland at Port Austin.

Charity Island Light (1857)

Big Charity is a 280-acre island located at the approach to Saginaw Bay. Along with its five-acre sister, Little Charity Island, it was named by early fishermen who believed the two islands were placed there by God's charity as a refuge from storms. The Lighthouse Board erected a conical tower with an attached keeper's dwelling in 1857 at a cost of $4,819. After 1900, when an acetylene lamp replaced one using kerosene, the resident keeper was no longer needed. This site became one of the earliest examples of lighthouse automation. The light was discontinued in 1939 when the Lighthouse Service built a new light at Gravelly Shoal about a mile west of Big Charity Island. The newer light, a modern square steel tower on a round concrete pier, remains in

service. The Federal government sold the old lighthouse and grounds to private interests in 1963. Today, the conical brick tower and large attached two-family keeper's house show the effects of decades of exposure to the elements.

Saginaw River Rear Range Light (1876, 1919)

The first lighthouse at the mouth of the Saginaw River was built after Congress received several memorials (petitions) from Saginaw County residents complaining about the dangers of an unlighted river entrance. One memorial dated October 29, 1834 contained a total of 112 signatures, and a year later, 63 individuals signed a second petition. Congress appropriated $5,000 in 1837, and approved a similar amount the following year, but the light did not go into service until 1841. The structure now standing replaced the original lighthouse, which was badly deteriorated by the 1870s. Congress then

appropriated $23,000 for a pair of range lights which went into service on September 15, 1876. The station consists of a two-story brick keeper's dwelling, 26 feet, 6 inches square, with a hipped roof, and a brick light tower measuring 15 feet square and attached to the northwest corner of the dwelling. The tower is 77 feet high, from the base to the top of the 10-sided cast iron lantern, which no longer displays a light. In 1919, a new reinforced concrete walkway five feet in width replaced the original raised wooden walkway encircling the keeper's dwelling.

Tawas Point Light (1876)

A light station has existed at Tawas Point (Ottawa Point) since 1853, but by the early 1870s, the exact location of the point had actually moved more than a mile because of shifting sands, thus necessitating the relocation of the lighthouse. Congress appropriated $30,000 for this purpose in March, 1875 and the work was completed the following year. The conical brick light tower is 67 feet high from the base to the top of the ventilator ball and creates a lens focal plane 70 feet above the mean low water level of Lake Huron. The tower is 16 feet in diameter at the base, 9 feet, 6 inches in diameter at the parapet, and is surmounted by a 10-sided cast iron lantern. At the base of the tower, there is an outer wall of brick 22 inches thick that covers an

air space of 24 inches, and an inner brick wall 8 inches thick. The original lens, no longer in place, was a fixed Fourth Order Fresnel, with a flash produced by a hood occultating device driven by clockwork. The lens which is now exhibited is a Fourth Order Fresnel bearing the inscription, "Barbier & Fenestre, Paris, 1880," with an electric motor providing rotation. A brick passageway connects the tower with the rectangular keeper's dwelling, which is also made of brick, measuring 26 feet by 43 feet, and topped by a gabled roof.

Sturgeon Point Light (1869)

Congress appropriated $15,000 in March, 1867 for the construction of a lighthouse on Sturgeon Point, which juts out into Lake Huron, thus creating a hazard to navigation. The station was finished in 1869 and the light was ready to be exhibited in November, but the beacon was not actually put into service until the spring of 1870. The first keeper, Perley Silverthorn, had donated the land on which it was built. The original lens was replaced in 1887 by a Third-and-a-half Order Fresnel lens manufactured by Henri Le Paute of Paris. The conical brick tower is 70 feet, 9 inches tall from the base to the top of the ventilator ball and creates a lens focal plane 69 feet above the lake level. The tower is 16 feet in diameter at the base, where the walls are 4 feet, 6 inches thick, and 10 feet in diameter at the parapet,

where the walls are 18 inches thick. It is surmounted by a 10-sided cast iron lantern with an inscribed diameter of 7 feet. A covered brick passage links the tower to the keeper's dwelling, a rectangular brick building with a gabled roof. The site is now a museum operated by the Alcona County Historical Society.

Alpena Light (1914)

The first lighthouse at the mouth of the Thunder Bay River in Alpena was built in 1875. It was originally a wooden tower, and this structure was later rebuilt in 1888. The skeletal steel-framed tower now standing dates from 1914. It consists of four steel corner posts resting on concrete piers three feet square, with the posts 12 feet, 6 inches apart. This pyramidal structure supports a round steel watchroom 12 feet in diameter, surmounted by a 10-sided cast iron lantern with an inscribed diameter of 10 feet. The tower creates a focal

plane 64 feet above the water level of the lake for the Fourth Order Fresnel lens which is still in service.

Thunder Bay Island Light (1832, 1857, 1868, 1893)

This light station, one of the earliest on Lake Huron, was established in 1832. Jesse Muncy was one of its first keepers. The original light tower was constructed of stone and stood 40 feet high to the parapet, with a diameter of 21 feet at the base and 11 feet, 4 inches at the top. The tower was raised ten feet in height in 1857, producing the structure now standing. This renovation was accomplished by encasing the upper fourteen feet of the old tower with a new brick wall and then extending the brickwork to the higher elevation. The surviving tower is 50 feet high to the parapet and is surmounted by a 10-sided cast iron lantern. Overall, it produced a lens focal plane 63 feet above the mean low water level of Lake Huron. The walls at the base of the tower are 6

88

MIDDLE ISLAND LIGHT

feet, 7 inches thick, but then taper to a thickness of only 20 inches at the parapet. By 1858, this light station had a fog bell struck by machinery, but in 1893, a fog signal building was added. The original steam-driven fog whistles are no longer present, having been replaced by a pair of diesel engines driving Ingersoll-Rand air compressors. The original lens in use here was a Fourth Order Fresnel with six flash panels, manufactured by Sautter of Paris, but it was also replaced some time ago by a modern airport-type beacon. The lightkeeper's dwelling, linked to the tower by an enclosed passage, is a two-story rectangular yellow brick building with a gabled roof, measuring 28 feet by 43 feet. This residence was reconstructed in 1868.

Middle Island Light (1905)

This light marks a major island located halfway between Thunder Bay Island and Presque Isle on the

OLD PRESQUE ISLE LIGHT

PRESQUE ISLE LIGHT

92

INTERIOR VIEW OF THE LANTERN AT PRESQUE ISLE

PRESQUE ISLE FRONT RANGE LIGHT

Lake Huron coastal shipping lanes. The conical brick tower, white with an orange band in the middle, stands 71 feet tall. It originally had a 10-sided lantern housing a Fourth Order Fresnel lens, with a fixed red character. Today, the tower and the attached brick keeper's house are in considerable disrepair, having been abandoned by the Coast Guard long ago.

Old Presque Isle Light (1840)

Presque Isle, which means "almost an island" in French, is a small peninsula which has provided mariners with one of the best harbors of refuge between Port Huron and Michilimackinac. Large numbers of ships engaged in coastal trade began using the harbor for shelter and as a source of cordwood for their boilers in the 1830s, prompting Congress to appropriate funds for a lighthouse there in July, 1838. The Federal Government requested bids a year later and Jeremiah Moors of Detroit, a mason and builder, won the contract. The lighthouse was completed in September, 1840 and Henry L. Woolsey was appointed its first keeper. Woolsey served through September, 1849, when he was replaced by Stephen V. Thornton, who served until September, 1855. His replacement, Louis J. Metevier, held the post until the end of September, 1861, when the last keeper at this station, Patrick Garraty, assumed the post. In 1868, the Lighthouse

Board decided to close this facility and build a taller coastal light nearby. When the new lighthouse opened in June, 1871, Garraty simply moved to the new station. The Federal Government sold the Old Presque Isle Light to E.O. Avery in 1897, but the Stebbins family has owned the property since the turn of the century and now opens the lighthouse to the public during the summer months. The conical tower is 30 feet tall, 18 feet in diameter at the base, and 9 feet in diameter at the top. The bottom two-thirds was built of stone, while the top third is brick. The original octagonal cast iron lantern, along with the lens, are no longer present.

Presque Isle Light (1871)

This light was built as a replacement for the original Presque Isle Lighthouse (1840) located about one mile to the south. Construction on this light was begun in 1870 and was completed the following year, with the new light first exhibited at the beginning of the 1871 shipping season. Patrick Garraty, the keeper at the old light, served at the new light until 1885, when his son Thomas succeeded him, serving until his retirement in 1935. The conical brick tower stands 113 feet high and produces a lens focal plane of 123 feet above the level of Lake Huron. The original lens, a Third Order Fresnel made by Henri Le Paute of Paris, remains in the lantern. The rectangular brick keeper's dwelling has a gabled roof.

PRESQUE ISLE REAR RANGE LIGHT

Presque Isle Range Lights (1870)

Congress appropriated $7,500 in March, 1869 for the construction of range lights to mark the channel into this important harbor and the project was completed the following year. The front light was housed atop a simple octagonal wood frame structure about 15 feet tall, while the rear light, located some 800 feet away, was exhibited some 36 feet above the lake level, in a lantern room built atop the keeper's dwelling. The rear light is a wood-framed rectangular building with a gabled roof, and it still has the tiny front porch where Anna Garraty maintained her all-night watches from her rocking chair during her long tenure as keeper.

Forty Mile Point Light (1897)

Beginning in 1890, the Lighthouse Board proposed the construction of a coast light halfway between existing lights at Presque Isle and Cheboygan because this fifty mile stretch of unlighted coast was a threat to navigation. Congress appropriated $25,000 for this station in August, 1894, and work began a few months later. Construction was completed by November, 1896, but the new light was not exhibited until May 1, 1897. Edward J. Lane was the assistant keeper during the first five years of operation. The station includes a double keeper's dwelling with an attached square brick light tower. The rectangular brick residence is 35 feet by 57 feet and has a gabled roof. The light tower, 12 feet square and 52 feet high, is surmounted by an octagonal cast iron lantern. The original lens, with a focal plane 66 feet above the water level of lake Huron, was a Fourth Order Fresnel with six bull's-eye panels, manufactured by Sautter of Paris. A different Fourth Order Fresnel lens, made by Henri Le Paute, is now exhibited. This station also includes a fog signal building, two surviving privies, and an oil house, all made of brick. □

94

FORTY MILE POINT LIGHT

6 THE STRAITS OF MACKINAC AND VICINITY LIGHTS

The Straits of Mackinac became an important battleground in the struggles between the French, British, and Americans for control of the upper Great Lakes, but since the early nineteenth century, this area has been a busy and dangerous bottleneck for shipping between Lake Michigan and the other Lakes. After Champlain discovered Lake Huron in 1615, the French explored the northern end of the Lake hoping to find the "Northwest Passage," the elusive all-water route to the Orient. Jean Nicolet, traveling in a simple birch bark canoe with a half dozen Huron Indians manning the paddles, passed through the Straits in July, 1634, and then continued along the northern shore of Lake Michigan. The strategic importance of this area was reflected in the two forts built here, one on Mackinac Island and the other on the south side of the Straits, just west of Mackinaw City. The many islands, reefs, and shoals in or near the Straits made this a dangerous passage for ships. Many of the earliest light stations on the Great Lakes were built here, including a lightship at Waugoshance Island (1832) and lighthouses at Bois Blanc Island (1829), South Manitou Island (1839), DeTour (1847), Skilligalee (1850), and Waugoshance (1851). This area is broadly defined here because the dangerous narrow passages which mariners had to navigate included the approaches to the Straits, which begin well south of Bois Blanc Island in Lake Huron and south of the Manitou Islands in Lake Michigan.

Cheboygan River Range Front Light (1880)

Congress appropriated $10,000 for the construction of range lights for Cheboygan Harbor in July, 1876, but it was not completed until the end of September, 1880. When built, the front range light had a Sixth Order Fresnel lens manufactured by Le Paute of Paris, but it now exhibits a pair of locomotive-type headlamps. The facility consists of a rectangular two-story frame keeper's house with a gabled roof, measuring 24 feet by 27 feet, 6

inches. The light tower attached to the dwelling, also of wood frame construction, is 9 feet square. It is topped off by a wooden lantern measuring 6 feet, 2 inches square, with a hipped roof. Keepers who served here included Ivory Littlefield (1883-1896), John Sinclair (1896-99), and John Duffy (1899-1902). This light should not be confused with the much older Cheboygan Light (1853), located about three miles east of the Cheboygan River, on Lighthouse Point. That light was abandoned by the Lighthouse Service in the 1920s, and now lies in ruins.

Fourteen Foot Shoal Light (1930)

This light was built in 1930 to mark this dangerous shoal near the entrance to Cheboygan Harbor. It consists of a concrete pier 50 feet square and 15 feet high resting on a rock-filled timber crib. The steel-framed tower is rectangular, 28 feet by 34 feet, and it is encased in one-quarter inch steel plates. It has a hipped roof and is surmounted by a cylindrical steel light tower which is 6 feet in diameter and extends 24 feet above the roof line. The 10-sided cast iron lantern originally held a Fourth Order Fresnel lens, but the lantern now exhibits a 250 millimeter plastic lens, with a focal plane 55 feet above the mean low water level of Lake Huron. This station was never manned, but was operated by radio-control from Poe Reef. This arrangement was an early experiment in off-site operations.

Poe Reef Light (1929)

This light station was built to replace the last of four different lightships which had warned ships of this dangerous reef since 1893. Poe Reef is located in the Straits of Mackinac South Channel, the route used by vessels passing between Lake Huron and Lake Michigan. The tower, which rests on a concrete pier 64 feet square and 45 feet high, is a steel-framed structure 25 feet square and 38 feet tall, surmounted by a square

SOUTH MANITOU ISLAND LIGHT

watchroom and a 10-sided cast iron lantern. It exhibits a Fifth Order Fresnel lens which produces a focal plane 71 feet above Lake Huron. Structurally, this station is identical to the light at Martin Reef, built in 1927.

Bois Blanc Island Light (1868)

Bois Blanc Island and the reefs nearby are located east of the Straits of Mackinac and were a major hazard to navigation by the early nineteenth century. In December, 1825, the merchants of Michilimackinac petitioned Congress for improved aids to navigation in the area and on May 23, 1828, Congress appropriated $5,000 to build a lighthouse on Bois Blanc Island. The contractor, Philo Scoville of Cleveland, completed the tower and keeper's house in 1829, making this station the second lighthouse on Lake Huron. The original tower collapsed in 1838, but a new tower was built the same year. Thirty years later, an entirely new tower and dwelling were built and these have survived to this day. Early keepers included Eber Ward (1829-42), Lyman Granger (1845-54), Mrs. Charles O'Malley (1854-55), Henry W. Granger (1855-57), and Mrs. Mary Granger (1857-?). The surviving buildings, sold by the Coast Guard to private interests in 1956, include a square yellow brick keeper's house with a gabled roof and an attached square light tower of the same materials. A yellow brick outhouse still stands as well.

Spectacle Reef Light (1870, 1874)

Spectacle Reef is really a pair of shoals where the water depth ranges from seven to eleven feet. But over the years, these shallow waters have caused the wreck of many vessels approaching the Straits of Mackinac. After two large schooners went down in these waters in the fall of 1867, Congress appropriated $100,000 in March, 1869, to begin construction of a lighthouse. A total of $406,000 was spent to complete the project, by far the most expensive lighthouse built on the Great Lakes up to that time. When finished, it stood 86 feet high, the tallest and most impressive example of monolithic stone lighthouse construction on the Lakes. Major O.M. Poe and Major Godfrey Weitzel, both of the U.S. Army Corps of Engineers, supervised the design and construction of Spectacle Reef Light, which was built with limestone brought from Marblehead, Ohio. The need for underwater foundation, the difficult climate of northern Lake Huron, and the isolated site combined to make this project a major engineering feat. Scammon's Harbor at Les Cheneaux, some sixteen miles distant from the reef, served as a base for materials and supplies. Over a four year period, a workforce of more than two hundred men, using two lighthouse tenders and a dozen other vessels, struggled during the shipping season to erect this structure. They built an outer cofferdam and then a smaller protective pier around the light tower, which

Poe Reef Light

Cheboygan Crib Light

Fourteen Foot Shoal Light

99

BOIS BLANC ISLAND LIGHT

was constructed of massive interlocking pieces of stone, all resting on bedrock well below the lake surface. After numerous battles with the elements, the task force finished the project in the spring of 1874 and displayed the permanent light for the first time on June 1, 1874. Early chief keepers at Spectacle Reef included William Marshall (1881-96), Samuel F. Rogers (1896-98), and Walter B. Marshall (1898-1902). This light station was automated in the early 1970s and has been unmanned since then. The Second Order Fresnel lens, made by Le Paute and installed in 1874, was removed in October, 1982 and is now on display at the Great Lakes Historical Society Museum in Vermilion, Ohio.

Round Island Light (1895)

Prior to the construction of this lighthouse, ships passing through the Straits of Mackinac from the Lake Huron side, including those coming from the St. Mary's

River, did not use the channel between Mackinac Island and Round Island because of many dangerous shoals in the vicinity. Instead, those vessels used the longer, but well-marked south channel between Bois Blanc Island and the mainland. After Congress appropriated the funds, a new lighthouse was built on Round Island in 1894-95. A keeper and two assistants operated the light until 1924, when automation reduced the staff to one. The lighthouse was entirely abandoned in 1947, and in 1958, the Coast Guard transferred the property to the United States Forest Service as an addition to the Round Island Scenic Area. In the early 1970s, high lake waters washed away the breakwater at the base of the square brick keeper's dwelling, opening up a hole in the foundation of the structure. Private historical societies, along with State and Federal agencies, cooperated in a full restoration of the station, with the main work completed by 1978.

Martin Reef Light (1927)

This light station marks a dangerous reef east of the Straits, and along the heavily used shipping route connecting Lake Michigan with Lake Superior via the St. Mary's River. A concrete pier, 60 feet square and 25 feet high, supports the steel-framed tower encased in steel plate. The tower is 25 feet square and 38 feet high to the parapet. The watchroom, which adds ten more feet to the height of the structure, supports the octagonal lantern, and produces a lens focal plane 65 feet above the lake level. The lens is a third order Fresnel, manufactured by Sautter and Company of Paris.

DeTour Point Light (1931)

The village of DeTour marks the southern entrance to the St. Mary's River, the connecting route between Lake Huron and Lake Superior. There has been a light at DeTour, located on the mainland, since 1848. The

ROUND ISLAND LIGHT

station was rebuilt in 1861 and was similar in design to the Manitou Island Light on Lake Superior. Since then, the light has been refitted several times and in 1902, the Lighthouse Board proposed to make the DeTour Light more distinctive and recognizable by installing a new lens that would produce a fixed light varied by a flash. On March 7, 1907 Congress appropriated $4,000 for this new lens, which was installed and first exhibited on May 12, 1908. The lens, along with the lantern in which it was placed, are the only parts of the onshore light station that were retained when the present structure was built in 1931 at the end of the DeTour Reef about one mile offshore. The lens is a Third-and-a-half Order Fresnel with four flash panels, each with a bull's-eye and concentric circular prisms in the central drum. It was made by Barbier, Benard & Turenne of Paris. The lens rotates once every 40 seconds, and produces a 10 second interval between flashes. The original manual clockwork turning mechanism is present, but it is no longer used. The 1931 light station is a steel-framed structure that rests on a concrete pier 60 feet square and 20 feet high. The lower segment of the station is 31 feet square, and is surmounted by a 12 foot square tower, which supports the watchroom and lantern, both 10-sided with an inscribed diameter of 9 feet. Overall, the structure stands 63 feet tall from the base to the top of the ventilator ball and produces a lens focal plane of 74 feet above water level. Charles R. Jones, who began work as

a keeper in the U.S. Lighthouse Service in 1922, served at DeTour Point from 1940 through 1963, when he was one of the last civilian keepers on the Great Lakes.

Old Mackinac Point Light (1892)

The passage through the Straits of Mackinac has been marked by a lighthouse at McGulpin's Point since 1869, but that light could not be seen from all points in the Straits. Subsequently, a fog signal was built at Mackinac Point in 1890, and the present lighthouse was completed in October, 1892. This beacon was visible for sixteen miles and it remained in active service through 1957. The lighthouse was especially valuable to the car ferries that shuttled between Mackinaw City and St. Ignace, and to the Mackinac Transportation Company, which operated ferryboat service to Mackinac Island. However, with the opening of the Mackinac Bridge in 1957, mariners used the bridge lights as their guide, rendering this lighthouse obsolete. The lighthouse was taken out of service in December, 1957, and after being declared surplus property a year later, was acquired by the Mackinac Island State Park Commission in 1960. The surviving buildings include a round light tower made of brick, resting on a cut stone foundation. The tower, six feet in diameter and forty feet tall, is attached to a two-story rectangular brick keeper's dwelling. The site now serves as a maritime museum.

101

OLD MACKINAC POINT LIGHT

MARTIN REEF LIGHT

DeTour Point Light

102

McGulpin's Point Light (1869)

Congress initially appropriated a sum of $6,000 on August 3, 1854, "for a lighthouse and fog bell at the south point of the harbor of Michilimackinac," but nothing was done, and more than a decade passed before the project was revived. Congress later approved another $20,000 on July 28, 1866, to build a new lighthouse, which was completed in 1869 and remains largely intact today. James Davenport, who had worked as a keeper at Waugoshance and at Little Sable Point in the early 1870s, served at McGulpin's Point from September, 1879 through December, 1906, when his position was abolished and the station was taken out of service. Davenport later worked at the Mission Point Light before his retirement in 1917. McGulpin's Point Lighthouse is a rectangular, yellow brick building with a gabled roof and an octagonal brick tower attached to one corner of the house. It has been a private residence for many decades.

Waugoshance Light (1851, 1870, 1883)

Waugoshance Shoal is a treacherous formation of islets that extend seven miles into Lake Michigan from Waugoshance Point, north of Sturgeon Bay. Because the water depth here is often twelve feet or less, this shoal was a major threat to navigation passing through the Straits of Mackinac. Consequently, a wooden lightship was stationed there from 1832 until 1851, when the first lighthouse was built. The new station rested on the first major crib structure built for a lighthouse on the Great Lakes. The timber crib, measuring approximately 32 feet by 60 feet, was filled with stone and sunk to its foundation on the reef. The conical brick tower is 76 feet high, with a diameter of 20 feet at the base and 15 feet, 2 inches at the parapet, with walls 5 feet, 6 inches thick at the base and 2 feet, 1 inch thick at the top. It is surmounted by a "bird cage" lantern, a style sometimes used in the 1850s, but surviving examples are extremely rare because polygonal lanterns became the standard design for lights on the Great Lakes by the 1860s. This lighthouse exhibited a Fourth Order Fresnel lens. A fog signal was added in 1883, and the tower and adjoining keeper's dwelling were encased in iron plates to protect the deteriorating brick walls. The pier on which the light tower rests has also undergone several major alterations since 1851. It was reconstructed in 1867-70, when a pier measuring 48 feet by 66 feet was built of massive limestone blocks, each 4 feet by 8 feet by 5 feet and weighing 12 tons. This stone structure was enlarged to 80 feet by 90 feet and covered with concrete in 1896, but this enlarged pier has disintegrated since then. The lighthouse was abandoned in 1912 after a new light station had been established at White Shoal in 1910.

Skillagalee (Ile Aux Galets) Light (1888)

A lighthouse was first established on Ile Aux Galets (Island of Pebbles) in 1851, but for virtually all of its existence, it was officially known as Skillagalee Island Light. The first lighthouse was rebuilt in 1868, when it was fitted with a Third Order lens. Changes in the water level of Lake Michigan have alternately enlarged or reduced the size of this island, but because of its location near the shipping channel west of Waugoshance Island near the Straits of Mackinac, it has remained a hazard to navigation. The constant erosion problems at this site, however, have also caused severe problems in protecting the lighthouses that have stood here. The present tower dates from 1888 and was the third major reconstruction of this facility. The 1868 tower was demolished in the spring of 1888, and a new tower was completed by the following October at a cost of $6,875. The keeper's dwelling and fog signal building that once stood here were demolished in 1969, so only the light tower remains. It is an octagonal brick structure 58 feet high, 14 feet in diameter at the base, and 9 feet in diameter at the parapet. The octagonal cast iron lantern, with an inscribed diameter of 7 feet, 1 inch, exhibits a Fourth Order Fresnel lens made in 1886 by Barbier & Fenestre of Paris. The tower creates a focal plane 58 feet above the mean low water level of Lake Michigan.

White Shoal Light (1910)

White Shoal, located twenty miles due west of Old Mackinac Point at the north end of Lake Michigan, has been a significant threat to navigation through the Straits since the later part of the nineteenth century. One of the first three lightships placed in service on the Great Lakes in the fall of 1891 was assigned to White Shoal. Because of the difficulty in moving lightships into place early in the shipping season, as well as removing them at season's end at remote locations such as this,

permanent light stations were preferred. On March 4, 1907, Congress approved $250,000 for the construction of a light station at White Shoal to replace the lightship. Construction began in 1908 and the new light was first exhibited on September 1, 1910. It was designed and built under the direction of Major William V. Judson, an engineer with the U.S. Army Corps of Engineers who was also the lighthouse engineer for the Lake Michigan district. The Edward Gillen Dredge & Dock Company of Racine, Wisconsin built the substructure for the light. Construction of the timber crib and concrete pier supporting the light tower was difficult because of the isolated location of the site, some 25 miles from the nearest protected harbor. The site was made level by placing a layer of stone riprap 102 feet square on the gravel lake bottom. The massive timber crib, 72 feet square and 18 feet, 6 inches high, was built at St. Ignace and then towed 28 miles to the site. After being sunk, the crib was filled with stone brought by a scow from

SKILLAGALEE (ILE AUX GALETS) LIGHT

Chicago, some 300 miles away. The concrete pier, 70 feet square and 22 feet high, was then built on top of the crib, which lay two feet under the water level. The bottom four feet of the pier was made of prefabricated concrete blocks and concrete mixed at the site, while the rest of the pier was made entirely of ready-mixed concrete. The construction of the substructure, including the pier, was all completed in the 1908 shipping season, which featured unusually mild weather. The conical light tower has a steel skeletal frame and is lined with brick and faced with dark enamel terra cotta blocks. The tower is 42 feet in diameter at the base, 20 feet in diameter at the parapet, and supports a round watchroom and lantern 12 feet, 6 inches in diameter. Overall, the tower is 121 feet tall from the base to the top of the ventilator ball and creates a lens focal plane 125 feet above the level of Lake Michigan. The lens was a Second Order Fresnel, with two panels each consisting of a 7-inch bull's-eye lens surrounded by

6 concentric prisms, that were covered by 15 distinct prism panels. The lens revolved once every 16 seconds on a mercury float trough, and was rated at 360,000 candlepower. It had a nominal range of twenty-eight miles. The lens bears the inscription, ''Phares & Fanaux, Barbier, Benard & Turenne, 82 Rue Curial, Paris.'' The original structure remains largely unchanged, except that a layer of Gunite has been applied to the exterior of the tower and a red ''barber pole'' stripe has been added to make the light more distinctive. The lens was removed in September, 1983, and the following December, George Keller, who had served as keeper at White Shoal from 1921 until 1946 passed away at age 85.

Gray's Reef Light (1936)

Gray's Reef, located west of the Straits of Mackinac near the main shipping channel, was considered such a

serious threat to navigation that it received one of the earliest Federal lightships placed in service on the Great Lakes. No. 57 assumed its station in October, 1891 and remained there until the new tower was built in 1936. Identical to the structure at Minneapolis Shoal, also in Lake Michigan, the light tower has a steel-framed base measuring 30 feet square and 15 feet tall, surmounted by a steel tower 17 feet tall to the parapet, which is 16 feet square at the base, and tapers to 10 feet square at the top. The entire structure is sheathed in steel plates, and rests on a steel and concrete pier which is 64 feet square and 30 feet high. The Fourth Order Fresnel lens is encased in a round steel lantern 8 feet in diamater.

St. Helena Island Light (1873)

St. Helena Island has long been a major center for the fishing industry in this region and was the site of several battles between the Mormons and the ''gentiles'' in the early 1850s. Congress appropriated $14,000 for a lighthouse in June 10, 1872 and it was built the following year. Its keepers included Thomas Dunn (1873-1875), Charles Marshall (1888-1900), Joseph Fountain or LaFountaine (1901-1920), and Wallace Hall, its last keeper, who left this station in 1923. At that time an acetylene lamp was installed the station was automated. The method of automation was unusual. Metal rods were installed that would expand as they were warmed by the sun, thereby shutting off the fuel supply. Around dusk, the rods would begin to contract, so the fuel could once again reach the lantern, thus allowing the light to burn brightly all night long. The original Third-and-a-half Order Fresnel lens made by Le Paute remained in place until recently, when it was replaced by a 300 millimeter plastic lens. The brick tower, 71 feet tall, is attached to a two-story, rectangular brick keeper's dwelling, measuring 26 by 31 feet, with a gabled roof. These structures are now in very poor condition.

Lansing Shoal Light (1928)

Lansing Shoal, another major threat to navigation on Lake Michigan, had been marked by a gas buoy for several years when Lightship No. 55 went into service there in July, 1900. However, beginning in 1908, the Lighthouse Board recommended building a permanent light station to replace the lightship, which often could not arrive there early enough in the shipping season or remain there late enough at season's end to guarantee mariner safety. It was not until 1928 that a permanent light station was erected there, one of the last major new offshore light stations constructed on the Great lakes. The lighthouse rests on a concrete pier 74 feet square and 20 feet high, and the pier sits on a timber crib 21 feet high. The steel-framed structure, encased in steel plate, has a segment 32 feet square and 12 feet, 8 inches high surmounted by the tower, which is 13 feet square at the base, 11 feet square at the parapet, and 30 feet, 4 inches

high. The circular cast iron lantern with helical bars across the glass panels exhibits a fixed Third Order Fresnel lens. Overall, the lighthouse structure is 59 feet high measured from the base to the top of the ventilator ball and creates a lens focal plane 69 feet above the mean low water level of Lake Michigan.

Seul Choix Point Light (1892, 1895, 1925)

Seul Choix Point marks a tiny harbor located on the south shore of Michigan's Upper Peninsula, some sixty miles west of the Straits. It means "only choice," and it was given this name by the French who found it to be the only harbor of refuge in this part of Lake Michigan. Later, it also provided Great Lakes shipping with protected anchorage during storms, but it remained a dangerous site until it was marked by a light. Congress appropriated $15,000 to construct a light and fog signal there in August, 1886, but the work proceeded very

BEAVER ISLAND (BEAVER HEAD) LIGHT

slowly. There were problems in acquiring land and the initial bids submitted by contractors exceeded the funds available, so an additional $3,500 had to be approved for the project. The light was placed into service in 1892, but the tower had to be rebuilt and the station was not entirely completed until September, 1895. The conical brick light tower rests on an ashlar foundation 12 feet high, with 5 feet below grade, and has a diameter of 18 feet at the base of the brickwork, and 12 feet, 8 inches at the parapet. The tower is surmounted by a 10-sided cast iron lantern that originally held a Third Order Fresnel lens manufactured by Le Paute of Paris. The lantern is now fitted with a modern airport beacon lens. Overall, the tower is 78 feet, 9 inches tall, measured to the top of the ventilator ball, producing a lens focal plane 80 feet above the mean low water level of Lake Michigan. The two-story brick keeper's dwelling, attached to the light tower by an enclosed walkway, measures 28 feet by 32 feet and has a gabled roof. It was modified in 1925 when an attached lean-to measuring 15 feet by 17 feet was replaced by a one-story rectangular brick addition, about 28 feet square. This alteration enabled the keeper's house to accommodate an additional family.

Squaw Island Light (1892)

Squaw Island is the northernmost island in the Beaver Island archipelago. The lighthouse has a red brick octagonal tower attached to a red brick keeper's dwelling. It exhibited a fixed red light, varied by a red flash every fifteen seconds, and its beacon was visible for thirteen miles. This station, abandoned long ago, also operated a fog signal which was a 10-inch steam whistle.

Beaver Island (Beaver Head) Light (1858, 1866)

Beaver Head, at the southern end of Beaver Island, marks the west side of the approach used by most vessels passing from Lake Michigan into the Straits of Mackinac. The first lighthouse was completed in 1852, but the tower was rebuilt in 1858 and the keeper's house was added in 1866. The first keeper was Mr. Loaney, a Canadian, who was followed by Captain Appleby of Buffalo, New York. Harrison Miller then served until 1905, when William Duclon took over, followed by Owen Gallagher. The Coast Guard constructed a radiobeacon at this site in 1962 and decommissioned the lighthouse, removing the illumination apparatus and lens. The Charlevoix Public School District subsequently bought the buildings and now uses them as an environmental education center. The surviving structures include a cylindrical yellow brick light tower with a 10-sided steel lantern room; a two-story rectangular yellow brick keeper's dwelling, 26 feet by 30 feet, a rectangular red brick fog signal building which is 22 feet by 40 feet, and other miscellaneous outbuildings.

110

Beaver Island Harbor (St. James) Light (1870)

The first lighthouse at this site was built in 1856, enabling mariners to use this harbor as a refuge from storms in the northern part of Lake Michigan. In 1867, the Lighthouse Board described the harbor as "indifferently lighted" and proposed the construction of a taller tower with a more powerful light. Congress appropriated $5,000 for this purpose in July, 1868, and the work was completed in 1870. The first keeper, Lyman Granger, was succeeded by Peter McKinley, who resigned his position in August, 1869, because of poor health. At that time, Elizabeth Whitney's husband, Mr. Van Riper, was appointed keeper. He later died trying to rescue the crew of the schooner *Thomas Howland* in Beaver Harbor and she was then named keeper. In 1884, she was appointed keeper at the Little Traverse Light in Harbor Springs. All that remains of the Beaver Island Harbor Light Station is the cylindrical brick

tower, 41 feet high, surmounted by a 10-sided cast iron lantern exhibiting the Fourth Order Fresnel lens installed in 1870. The lens bears the inscription, "Barbier & Fenestre, Crs., Paris."

South Fox Island Light (1868, 1934)

South Fox is one of several islands along the approach to the Straits of Mackinac from Lake Michigan. The first lighthouse here was completed in 1868 at a cost of $18,000 and it consisted of a rectangular brick keeper's house with an attached square brick light tower about thirty feet tall. Although these buildings are still standing, they are badly deteriorated. A later tower built closer to the shoreline in 1934 has survived as well, but it is also no longer used. This tower is a skeletal steel structure, 60 feet high, with an enclosed round stair tower made of steel plates that leads to the 10-sided lantern room at the top of this station.

SOUTH MANITOU ISLAND LIGHT

North Manitou Shoal Light (1935)

This light station is a steel-framed structure encased in steel plates, and similar in overall design and appearance to the Lansing Shoal Light constructed in 1928. It was built on a square concrete crib and replaced a light vessel which had marked this hazard to navigation since the second decade of this century.

South Manitou Island Light (1858, 1872)

South Manitou Island has been an important part of the Lake Michigan maritime system from the 1830s on. It was an excellent source of hardwood fuel for the lake steamers, and was one of the few deep natural harbors between Chicago and the Straits. It is also situated on the heavily traveled Manitou Passage, the most important route to the Straits of Mackinac. The Federal government built the first lighthouse here in 1839 and Dr. Alonzo Slyfield served as keeper from 1848 to 1859. The surviving brick keeper's dwelling dates from 1858, when the entire station was rebuilt. The 35 foot tower erected at that time created a lens focal plane 64 feet above the water level, but this height was quickly deemed inadequate by the Lighthouse Service. The new tower, completed in 1870, is a conical brick structure 18 feet in diameter at the base and 104 feet tall, creating a lens focal plane 100 feet above the lake level. A Third Order Fresnel lens was installed, giving the light a nominal range of eighteen miles. A large steam fog signal was added in 1875, replacing the fog bell that had been in service here for many years. Later keepers such as Martin Knudsen (1881-1889) viewed South Manitou Island Light as an attractive assignment and it remained an important light station well into the twentieth century. In 1958, the Coast Guard abandoned the station and it is now a historical museum which is part of the Sleeping Bear Dunes National Lakeshore. □

VIEW FROM THE TOP OF THE TOWER AT SOUTH MANITOU

7 THE LIGHTS OF LAKE MICHIGAN

Jean Nicolet discovered Lake Michigan in 1634 as part of the continuing effort by the French to discover a northwest passage to the Pacific Ocean. The inland sea he discovered was named "Lake of the Stinking Water" for many years. After the explorations of Marquette and Jolliet in 1673-75 delineated its boundaries in detail, cartographers called it Lake Michigan or Lake Illinois. Because of Lake Michigan's large size and strategic location, it remained an important prize in the battles between the Indians, French, British, and Americans for control of the upper midwest. Not until 1815, after the conclusion of the War of 1812, did Lake Michigan permanently fall under American control.

The economic development of the lands bordering the lake came quickly. The *Walk-in-the-Water,* the first steamship on the Lakes, ventured onto Lake Michigan in 1821 with a load of troops destined for Green Bay. The Erie Canal opened in 1825 and in 1830, regular passenger service began between Buffalo and Green Bay. In the early 1830s, the partnership of Oliver Newberry and George W. Dole began exporting meat and wheat from Chicago to Detroit and points east, launching Chicago's phenomenal growth. By the time of the "Great Fire" in October, 1871, Chicago had seventeen grain elevators with a combined capacity of over eleven million bushels. The enormous grain trade with the east coast, combined with the bulk shipments of Lake Superior iron ore to the steel mills of Gary and vicinity, gave Lake Michigan the largest volume of traffic of all the Lakes by the end of the nineteenth century. Subsequently, it also has had the most lighthouses from the 1850s to the present.

Little Traverse (Harbor Point) Light (1884)

This light station was established in 1884, with the light first exhibited on September 25. Mrs. Elizabeth Whitney Williams was the first keeper, after previously serving at the Beaver Island Harbor Light. The surviving buildings include a two-story red brick keeper's house with an attached square red brick light tower. A Fourth Order Fresnel lens with the markings, "Sautter, Lemonnier," is still in place. A rare example of a fog bell has also survived.

Charlevoix South Pier Light (1948)

The first pier light at Charlevoix, a square wooden structure dating from 1885, was originally located on the north pier, but was moved to the south pier in 1914, where it was replaced by the present light in 1948. It is a skeletal steel tower, exposed at the bottom, but enclosed at the top with steel plates. The 10-sided lantern holds a Fifth Order Fresnel lens.

Mission Point (Old Mission Point) Light (1870)

The Peter Dougherty Mission, established in 1839 at the end of the peninsula, dividing Grand Traverse Bay gave this location its name. Congress initially appropriated a sum of $6,000 for a lighthouse in 1859, but the Civil War and other problems delayed construction until 1870. The original frame dwelling, with a square tower projecting from the gabled roof of the house, has survived, but without the original lighting apparatus. Keepers who served at Mission Point include Jerome Pratt (1870-77), Jonathan McHarry (1877-81), Jonathan Lane (1881-1906), Mrs. Sarah E. Lane (1906-08), and James Davenport (1908-1919). Emil Johnson, the last keeper at Mission Point, left after the station closed in June, 1933. The State of Michigan purchased the property after the Second World War and created a park there. The lighthouse is maintained by Peninsula Township and is used as a residence by township employees. The Mission Point Light is located almost exactly on the 45th parallel of latitude, which is halfway between the Equator and the North Pole.

SOUTH HAVEN SOUTH PIER LIGHT

115

Grand Traverse (Cat's Head Point) Light (1858, 1899)

Congress appropriated $4,000 on September 28, 1850 to establish a lighthouse at Cat's Head Point to mark the entrance into Grand Traverse Bay. It was completed in 1853 and the first Keeper was Deputy U.S. Marshal Philo Beers, who served from April 1853 until September 1857 when his son Henry took the post. The lighthouse now standing was built in 1858 and consists of a rectangular brick keeper's house measuring 30 feet by 47 feet, a two-story building with a gabled roof. The light tower extends from the roof of the residence and is a square frame structure 7 feet, 6 inches on a side, surmounted by a 9-sided cast iron lantern with an inscribed diameter of 7 feet, 4 inches. The lens, originally a Fourth Order Fresnel made by Barbier and Fenestre, is no longer in place and this facility no longer serves as a lighthouse. A rectangular brick fog signal building constructed in 1899, and measuring 22 feet by 40 feet, stands nearby. A small rectangular brick oil house has also survived. The Coast Guard built a steel skeletal tower with an automatic light in 1972, when they abandoned the old lighthouse, removing the lens and lighting equipment in the process. The buildings and lands are now leased by the State of Michigan and form a portion of Leelanau State Park.

Point Betsie Light (1858, 1894)

Originally named ''Pointe Aux Becs Scies,'' meaning ''sawed beak point,'' this prominent point is located at the southern end of the Manitou Passage. In 1853, Congress authorized the expenditure of $3,000 for a lighthouse at this location, but it was not completed until 1858. The first keeper, Dr. Alonzo Slyfield, had previously served on South Manitou Island. He and his two sons served as keepers at Point Betsie for a total of twenty-two years. Later keepers include Medad Spencer

LITTLE TRAVERSE (HARBOR POINT) LIGHT

MISSION POINT (OLD MISSION POINT) LIGHT

(1894-1906), Phil Sheridan (1906-?), and Edward Wheaton (1934-46). This was such an important coastal light that the Lighthouse Board suggested in 1880 that the original 37 foot tower be replaced with a new 100-foot structure. The smaller tower, which has survived intact, was being seriously threatened by beach erosion, forcing the Lighthouse Board to take corrective action or risk losing the tower. Finally in April, 1890, they strengthened the tower foundation by underpinning it with a ring of concrete and at the same time they built a new revetment to reduce the erosion problems. The cylindrical light tower is surmounted by a 10-sided cast iron lantern and produces a lens focal plane 52 feet above the mean low water level of Lake Michigan. It contains a Fourth Order Fresnel lens, with flash panels, manufactured by Barbier & Fenestre. The two-story rectangular frame keeper's dwelling, with a gambrel roof, originally measured 28 feet square, but was enlarged in 1894 to present dimensions of 28 feet by 48 feet. The Coast Guard automated Point Betsie in the spring of 1983, when it was the last manned light station on the east shore of Lake Michigan.

Frankfort North Breakwater Light (1932)

There has been a lighthouse on the breakwater at Frankfort since 1873, but the structure standing there now was erected in 1932. It is a square steel-framed pyramidal tower covered in steel plate. There is a segment 15 feet, 6 inches square and 25 feet high that is surmounted by a pyramidal segment 14 feet, 6 inches square at its base and 10 feet square at the top, which in turn supports the 10-sided cast iron lantern with an inscribed diameter of 8 feet. Overall the structure is 67 feet high measured from the base to the top of the ventilator ball and creates a lens focal plane 72 feet above the mean low water level of Lake Michigan. It exhibits a Fifth Order Fresnel lens manufactured by Barbier, Benard & Turenne.

GRAND TRAVERSE (CAT'S HEAD POINT) LIGHT

Manistee North Pierhead Light (1927)

Congress authorized a light at the mouth of the Manistee River in 1860, but it was not built until 1870 and then the structure was almost immediately destroyed by fire in October, 1871. The replacement light was built on the south pier in 1873 and a new light was built on the north pier, where it remained until 1927, when the present light replaced it. The surviving station is a conical tower of riveted steel plate construction, 39 feet high, surmounted by a 10-sided lantern that is fitted with a 155 millimeter plastic lens.

*Big Sable Point (Grand Point Au Sable) Light
(1867, 1900, 1905)*

The Lighthouse Board noted in 1865 that Grand Point Au Sable was "the most salient point on the eastern shore of Lake Michigan, between Point Betsie

117

POINT BETSIE LIGHT

FRANKFORT NORTH BREAKWATER LIGHT

and Muskegon'' that remained unlighted and observed that ''the interests of commerce demand that it be suitably lighted.'' Congress appropriated funds for this project in July, 1866 and construction was completed the following year, with the light first exhibited on November 1, 1867. The conical brick tower was threatened by beach erosion within two years after it was built, but a longer-term problem was the deterioration of the brickwork. The tower was repointed in 1880, but the problem was not permanently solved until 1900, when the entire tower up to the watchroom was encased in steel plates, and the space between the plates and the brickwork was filled in with concrete. The watchroom was similarly encased in 1905. The reconstructed tower, which stands 112 feet to the top of the ventilator ball, is 19 feet, 2 inches in diameter at the base and 13 feet, 4 inches in diameter at the parapet and is surmounted by a round watchroom and a 10-sided cast iron lantern, with an inscribed diameter of 8 feet, 9 inches. Three of

the lantern panels facing the land are blocked. The lantern houses a Third Order fixed Fresnel lens bearing the inscription, '' Sautter & Co., Constructeurs.'' The light tower is connected by a covered passage to the keeper's dwelling, a rectangular framebuilding with a gabled roof, measuring 29 feet by 63 feet, built to house two keepers and their families.

Ludington North Pierhead Light (1924)

Private interests first undertook harbor improvements in the 1850s to permit access to the sawmills which had sprung up in this part of the state. In July, 1870, Congress authorized the spending of $6,000 for a ''beacon light at Pere Marquette Harbor, Lake Michigan, in the State of Michigan.'' The Pere Marquette River derived its name from the French missionary who died here, but the town was named after James Ludington, a lumber baron who developed the

MANISTEE NORTH PIERHEAD LIGHT

BIG SABLE POINT (GRAND POINT AU SABLE) LIGHT

LUDINGTON NORTH PIERHEAD LIGHT

LITTLE SABLE POINT LIGHT

area. When the first light went into service in 1871, the harbor was crowded with ships carrying lumber. In a 30-day period beginning May 15, 1871, sixty-seven vessels carrying nearly seven million feet of lumber cleared the harbor. The surviving structure, built in 1924, is a square pyramidal building, steel-framed and encased in riveted steel plates. It displays a glass lens bearing the inscription, "MacBeth Evans Glass Company, Pittsburgh."

Little Sable Point Light (1874)

Beginning in 1870, the Lighthouse Board urged the construction of a light station in this area because of the long stretch of coastline that was unlighted, and the increased volume of shipping traffic in the vicinity. Congress appropriated a sum of $35,000 for this light in June, 1872, and work commenced the following April. Because of the isolated location of the site, they had to construct docks for landing all the needed construction materials and supplies. The work was largely completed in 1873 and the light was exhibited in the spring of 1874. James Davenport was the first keeper. The original light, a three-wick oil-burning lamp, was replaced around 1918 with an incandescent oil vapor lamp using kerosene. After electric power lines finally reached this remote area, the Coast Guard installed an automatic electric light in 1954, and transferred the keeper, Henry Vavring, to Big Sable Point. They also demolished the keeper's dwelling, leaving only the tower. The conical red brick tower is 107 feet tall from the base to the top of the ventilator ball and produces a lens focal plane 108 feet above the mean low water level of Lake Michigan. The Third Order Fresnel lens, manufactured by Sautter and Co. of Paris, has a lower portion consisting of eight fixed panels, creating an arc of illumination of 300 degrees, and an upper segment, which consists of ten panels and rotates. The lantern is now driven by an

WHITE RIVER LIGHT

electric motor, but the original clockwork turning mechanism still remains in the tower.

White River Light (1875)

Congress first authorized the expenditure of $10,000 for a new lighthouse at this location in July, 1866, but no work was done until after a new channel was opened in 1870. A pier light costing $1,000 was built in 1871, but then no additional work was done until 1875-76, when the buildings now standing were completed. Captain William Robinson was the keeper from 1875 to 1915 and his son-in-law, William Bush, served from 1915 until the light station was closed in 1941. Fruitland Township later bought the property in 1966 and now uses the keeper's dwelling as a local museum. The octagonal yellow brick tower is attached to a one-story rectangular yellow brick residence. The original Fourth Order Fresnel lens is on display in the museum.

Muskegon South Pier Light (1903)

The first lighthouse at Muskegon was established in 1852, but prior to the construction of this light tower in 1903, the beacon was attached to the top of the keeper's dwelling on shore. At the turn of the century, however, the Lighthouse Board finally decided to erect a new steel tower on the end of the south pier and went ahead with construction in 1902. The conical tower was completed in 1903 and stands 48 feet high. The lens, a Fourth Order Fresnel, is semi-circular in form and is equipped with a brass reflector, producing an arc of illumination of 180 degrees. It bears the inscription, "Sautter & Cie, Constructeurs a Paris."

Grand Haven South Pier Inner Light (1905)

The harbor at the entrance to the Grand River from Lake Michigan was the only deep water harbor of refuge

MUSKEGON SOUTH PIER LIGHT

MUSKEGON SOUTH BREAKWATER LIGHT

AT LEFT, GRAND HAVEN SOUTH PIER INNER LIGHT

AT RIGHT, GRAND HAVEN SOUTH PIERHEAD LIGHT

ANOTHER VIEW OF GRAND HAVEN SOUTH PIER

at the south end of the lake, so it became an important port early in the nineteenth century. The first Federal lighthouse here, known as the Grand River Light, was built in 1839 on the south side of the river, but it was replaced in 1855 by a new lighthouse placed on the high bluff east of the shore, with the lens about 150 feet above the water level. The second light was discontinued in 1895 and replaced by a structure placed at the outer end of the south pier. The surviving conical light tower, of steel plate construction, is 39 feet tall from the base to the parapet, and 51 feet tall overall. The American Bridge Company fabricated and erected the tower in 1905. The Coast Guard later gave the original Sixth Order Fresnel lens to the City of Grand Haven and replaced it with a plastic lens.

Grand Haven South Pierhead Light (1875, 1922)

This structure is probably the original fog signal building constructed in 1875 and located at the end of the south pier. By 1905, however, after the pier had been extended several times, the fog building was some 600 feet from the end of the pier, so it was moved to that point, and a new steel tower was built in its place. The rectangular wood-framed building was sheathed with corrugated iron in 1922 to prevent deterioration of the wood. The octagonal lantern is fitted with a plastic lens.

Holland Harbor (Black Lake) Light (1907)

The first lighthouse at this location was a wooden structure erected in 1872 and tended by M. Van Regenmorter. The Lighthouse Board erected a keeper's dwelling with a skeletal light tower in 1907, but the tower was removed in 1936 and the lantern placed in a tower projecting from one of the twin gabled roofs of the dwelling. The building is covered with steel plates,

HOLLAND HARBOR (BLACK LAKE) LIGHT

which are painted red. The original Sixth Order Fresnel lens was given to the Netherlands Museum and a 250 millimeter plastic lens remains in use today.

Saugatuck (Kalamazoo River) Light (1859)

Congress first authorized a lighthouse at the mouth of the Kalamazoo River in March, 1837, and construction was completed in 1839, making it one of the earliest lights on Lake Michigan. The first light was washed out by erosion and was replaced in 1859 with the lighthouse which survived until it was demolished by a tornado in 1956. When the Lighthouse Service discontinued the station in 1915, it was equipped with a bi-prismatic Fresnel lens with an oil lamp providing the illumination. After the 1956 tornado, the lumber from the old keeper's house was salvaged and used to build a summer cottage, which survives to this day.

South Haven South Pier Light (1903)

South Haven has had a lighthouse since 1872, but the steel plate structure now standing was built in 1903, with the light first exhibited on November 13. It originally had a Fifth Order fixed Fresnel lens made by Barbier & Fenestre, but it now exhibits a Sixth Order glass lens lantern. The cylindrical steel tower has a diameter of 11 feet, 6 inches at the base and stands 36 feet tall overall.

LIGHT TOWER AT HOLLAND HARBOR

125

SUMMER COTTAGE BUILT WITH LUMBER FROM OLD SAUGATUCK LIGHT

St. Joseph North Pier Outer Light (1907)

The first lighthouse built at the mouth of the St. Joseph River was completed in 1832, making it one of the two earliest lights on Lake Michigan. Thomas Fitzgerald was the first keeper. The original round stone tower was replaced in 1859 by another lighthouse built on shore. This light was finally discontinued in 1907. Several pier lights have been built on the north pier since the 1880s. The north pier was extended 1,000 feet in 1906, necessitating the reconstruction of the two lights which had served as range lights as well as harbor lights. This light, also serving as the front range light, is a conical structure consisting of cast iron plates and a steel frame, measuring 8 feet, 3 inches in diameter at the base and 7 feet, 3 inches in diameter at the parapet. It is surmounted by a round watchroom and a 10-sided cast iron lantern, giving the structure a total height of 35 feet, 8 inches. It exhibits a Fifth Order fixed Fresnel lens comprised of two panels creating an arc of illumination of 180 degrees and two brass reflecting panels, manufactured by Barbier & Benard of Paris.

St. Joseph North Pier Inner Light (1907)

The inner light (rear range light) on the St. Joseph North Pier was established in 1898, but was rebuilt in 1907 after the pier had been extended 1,000 feet the year before. It consists of a steel-framed fog signal building encased in 3/8 inch cast iron plates, 26 feet square, with a hipped roof, surmounted by an octagonal light tower of similar design. The tower has a diameter of 10 feet, 6 inches and supports a round cast iron lantern with helical bars across the lantern panels. Overall, the structure is 57 feet, 8 inches tall, measured from the base to the top of the ventilator ball. When rebuilt in 1907, this lighthouse exhibited a Fourth Order lens made by Chance Brothers Company of Birmingham, England,

SOUTH HAVEN SOUTH PIER LIGHT

ST. JOSEPH NORTH PIER

ST. JOSEPH INNER LIGHT AT LEFT   OUTER LIGHT AT RIGHT

but the lens currently in use is a Fifth Order fixed Fresnel, with an arc of illumination of 270 degrees and a brass reflector panel, manufactured by Sautter & Company of Paris.

Michigan City Light (1858)

This well-preserved lighthouse was built in 1858 at a cost of $8,000. It replaced an earlier wooden tower and dwelling built in 1837. According to the story, the light tower was blown down during a violent storm, just after the keeper, Miss Harriet Colfax, had lit the lamp. Miss Colfax, one of the few female Great Lakes lightkeepers, served a long tenure at this station from 1853 to 1904. This lighthouse served as the major harbor beacon until 1904, when the light was moved to a new structure on the east pier. The old lighthouse then served as a keeper's residence until 1940, and it was finally abandoned by the Coast Guard in 1960. The Michigan City Historical Society now leases the property from the city and maintains the facility as a lighthouse museum. It is a handsome two-story rectangular building with a light tower and lantern (not the original) projecting from the rooftop. The stone foundations support yellow brick walls which make up the first floor, while the second floor has shingle siding. There is a Fourth Order Fresnel lens with the markings, ''Barbier, Benard & Turenne, Paris,'' on display in the museum.

A VIEW FROM THE TOWER AT ST. JOSEPH INNER LIGHT

Michigan City East Pier Light (1904)

The first lighthouse at Michigan City was constructed in 1837 and rebuilt in 1858, but beginning in 1894 the Lighthouse Board recommended that a fog signal be added to the station. Congress finally appropriated $5,500 for this purpose on June 6, 1900, but work on the new building was not begun until June, 1904, after the extension of the east pier was completed. In the meantime, new lamps were installed at the old lighthouse located on shore in 1902. The lantern, lens, and lamps were then moved into the new structure and were placed in service on October 20, 1904. The new structure, resting on a concrete foundation, consists of a steel framed building lined with brick and encased in steel plates, measuring 24 feet square and 15 feet high to the plate line. It has a hipped roof surmounted by an octagonal tower 10 feet, 6 inches in diameter supporting a round cast iron lantern with helical bars running across the glass panels. Overall, the structure stands 49 feet high from the base to the top of the ventilator ball and produces a lens focal plane 55 feet above the mean low water level of Lake Michigan. The lens is a Fifth Order Fresnel fitted with a brass reflector covering 90 degrees, and producing an arc of illumination of some 270 degrees. It was manufactured by Sautter and Co. of Paris. However, none of the original fog signal equipment remains in the structure.

NORTH PIER OUTER LIGHT COVERED WITH SNOW AND ICE

AT LEFT, MICHIGAN CITY BREAKWATER LIGHT

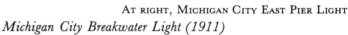

AT RIGHT, MICHIGAN CITY EAST PIER LIGHT

Michigan City Breakwater Light (1911)

This light was built in September, 1911, after the breakwater was rebuilt following its destruction in a storm in February, 1910. The light tower is a concrete pyramidal structure resting on a concrete base, that tapers from 5 feet square at the base to 2 feet, 6 inches square at the top, and is surmounted by a plastic lens lantern. Overall, the structure is 32 feet high and creates a lens focal plane 36 feet above the lake level.

Gary Breakwater Light (1911)

This cylindrical red steel-plate tower rests on a round concrete crib and creates a lens focal plane 40 feet above the lake level. A Sixth Order Fresnel lens remains in the original lantern as a possible auxiliary light for another lens lantern positioned on top of the tower.

Buffington Breakwater Light (1926)

The conical steel tower, painted red, rests on a round concrete crib and has a ten-sided lantern. The structure creates a lens focal plane 48 feet above the water level. This light, now maintained as a private aid, is a good example of standard pier light tower designs used in the the first few decades of this century.

MICHIGAN CITY LIGHT

130

GARY BREAKWATER LIGHT

BUFFINGTON BREAKWATER LIGHT

131

INDIANA HARBOR EAST BREAKWATER LIGHT

CALUMET HARBOR (SOUTH CHICAGO) LIGHT

Indiana Harbor East Breakwater Light (1935)

This steel square tower, painted white, rests on a square concrete base and stands 78 feet tall. It is fitted with a lens lantern, which is probably part of the original equipment installed at this station. This light is identical to the Port Washington, Wisconsin, Breakwater Light, also built in 1935.

Calumet Harbor (South Chicago) Light (1906)

The U.S. Bureau of Topographical Engineers completed the first lighthouse at the mouth of the Calumet River in 1851, but this station was entirely rebuilt in 1873 and again in 1876. The present structure is the fourth light at Calumet. Contracts for this lighthouse were let in 1904, and construction began in the fall of 1905. The completed light was first exhibited on July 20, 1906. It consists of a rectangular fog signal house, a steel-framed structure encased in steel plates, surmounted by a conical steel tower 10 feet, 6 inches in diameter at the base, with a round lantern 7 feet, 8 inches in diameter, with the helical bars across the glass panels. The original Fourth Order Fresnel lens is gone and the lighthouse now exhibits a modern plastic lens.

Chicago Harbor Southeast Guidewall Light (1938)

This structure is one of many new lights built in the modern era. It is a square steel tower, supported by four steel legs, open at the bottom, but with the upper half enclosed with steel plates, surmounted by an octagonal lantern encasing the lens.

Chicago Harbor Light (1893, 1919)

In 1831, Congress authorized the first lighthouse at Chicago, to be built at the mouth of the Chicago River. Completed in 1832, it was the earliest lighthouse on

CHICAGO HARBOR SOUTHEAST GUIDEWALL LIGHT

CHICAGO HARBOR LIGHT

Lake Michigan, along with one at St. Joseph. The Bureau of Topographical Engineers later supervised the building of a new lighthouse at the end of the north pier in Chicago Harbor and completed the work in 1852. When the new light went into service, the old one on the mainland was discontinued. The lights on the Chicago piers have been altered several times since the 1860s. In 1872-73, the Lighthouse Board moved the main light from the Chicago pier to Grosse Point, where it became the chief coastal light for this area. The first light on the north breakwater was constructed in 1876, but the surviving light, originally built in 1893, was located on the mainland at the entrance to the Chicago River. The Lighthouse Board had exhibited a Third Order lens consisting of red and white panels at the Columbian Exposition in Chicago in 1891, with the intention of installing the lens in its new lighthouse at Point Loma, California. The lens was still on display when the new Chicago Harbor Light was completed in 1893, so the Lighthouse Board installed it there instead. Later, Congress appropriated $88,000 in June, 1917 to move the entire lighthouse to the newly built south end of the north arm of the outer breakwater and this work was completed in 1918-1919. It is a brick-lined steel tower 18 feet in diameter and 48 feet high to the base of the watchroom, which is surmounted by 10-sided cast iron lantern with vertical bars. There is an attached fog signal building measuring 28 feet square located on one side of the tower and a rectangular boathouse measuring 18 feet by 28 feet on the other side, both with hipped roofs. The structure produces a focal plane of 82 feet above the mean lower water level.

Grosse Point Light (1873, 1880)

Named by the French fur traders of the eighteenth century, Grosse Point has served as a major landmark for Lake shipping serving Chicago. The nearby shoals became a major hazard as the volume of traffic rose sharply with Chicago's growth. By 1870, the city received more than a thousand vessels a month. Congress appropriated $35,000 in March, 1871, and an additional $15,000 in March, 1873, for the construction of a Second Order coastal light at Grosse Point. The light station included a conical brick tower, 90 feet tall, an adjoining duplex keeper's dwelling, and two brick fog signal buildings added in 1880. The Second Order Fresnel lens, made by Le Paute, was fitted with a three-wick kerosene lamp, and was rotated by a weight-driven clockwork mechanism to give the light a flashing characteristic. The lens focal plane was 121 feet above the lake and on a clear night the light could be seen for 21 miles. Because the walls of the light tower were deteriorating, they were covered with four inches of concrete in 1914. The steam fog signals installed in 1880 were removed in 1922 and the light was converted to

KENOSHA PIERHEAD LIGHT

WAUKEGAN HARBOR (LITTLE FORT) LIGHT

electricity and automated in 1935. The City of Evanston took over most of the station in 1935 and in 1941, received the tower and lighting apparatus as well. The light is still maintained as a private aid to navigation, but is primarily used as an historical museum. It is the only Second Order lens on the American side of the Lakes still in its original location atop a light tower.

Waukegan Harbor (Little Fort) Light (1889)

Congress appropriated a sum of $4,000 for a lighthouse at Little Fort, Illinois in 1847 and the structure was completed in 1849. Mr. D.O. Dickinson was the keeper here in the late 1850s. After a reconstruction in 1860, the light underwent no additional major changes until the late 1880s, when the surviving tower was built. When originally constructed in 1889, the tower was a conical cast iron structure measuring 21 feet in diameter at the base and standing 36 feet tall. It originally had a

KENOSHA (SOUTHPORT) LIGHT

135

WIND POINT (RACINE POINT) LIGHT

OLD RAILWAY LEADING FROM THE PIER AT WIND POINT

Fourth Order Fresnel lens. A fog signal building added in 1905 was later removed, and the lantern was also replaced by a plastic lens which is now in service.

Kenosha (Southport) Light (1866)

The first lighthouse at Southport (later renamed Kenosha), Wisconsin was constructed in 1848, but it was entirely rebuilt in 1866. The light station included a 40-foot conical brick tower, without a lantern, and a detached L-shaped two-story brick keeper's house, featuring a gabled roof and a large porch.

Kenosha Pierhead Light (1906)

There has been a lighthouse at the end of the Kenosha pier since 1864, but the surviving structure was erected in 1906. It is a conical tower consisting of riveted steel plates. The tower stands 50 feet tall from the base to the top of the ventilator ball and has a round lantern with helical bars across the glass panels. Originally equipped with a Fourth Order Fresnel lens, it now exhibits a modern rotating airport beacon.

Wind Point (Racine Point) Light (1880, 1900)

In 1870, the Lighthouse Board began recommending the construction of a lighthouse at Wind Point north of

Racine Harbor because this point could not be easily seen by ships coming into Racine from the north, plus it blocked the Racine lights from view. Congress finally appropriated funds for the project in 1877 and it was completed in 1880 at a cost of about $100,000. When the lighthouse was placed in service on November 15, 1880, it exhibited a Third Order Fresnel lens made by Barbier and Fenestre of Paris, providing a white flashing coastal light, but it also exhibited a Fifth Order red light which could be seen in the vicinity of Racine Reef, located to the south and marked only by a buoy. The lighthouse now exhibits a modern rotating airport-style beacon. The conical brick tower is 108 feet tall from the base to the top of the ventilator ball and is topped by a 10-sided cast iron lantern. The tower, resting on a stone foundation extending 10 feet deep to bedrock, is 22 feet in diameter at the base, where the walls are 7 feet thick. The tower is connected through a covered passageway to the keeper's house, a frame building with a gabled roof, measuring 70 feet by 28 feet overall. It contained three apartments to house the keeper and two assistants. The complex also included a brick fog signal building, erected in 1900, as well as several other structures.

Milwaukee Pierhead Light (1906)

The Milwaukee Pierhead Light was first established in 1872 and rebuilt in 1906. The tower that stands today

MILWAUKEE BREAKWATER LIGHT

is the 1906 reconstruction, but the lantern and lens have been altered. The conical tower measures 11 feet, 6 inches in diameter at the base and stands 42 feet high. It consists of riveted steel plates 3/8 inch thick. In 1910, it had a round lantern with helical bars housing a Fourth Order Fresnel lens manufactured by Sautter of Paris. Today, the lantern is 10-sided and contains a Fifth Order Fresnel lens made by Henri Le Paute of Paris, bearing a date of 1877.

Milwaukee Breakwater Light (1926)

The Milwaukee Breakwater Light, completed in 1926, is typical of the last generation of fully enclosed lighthouses erected on the Great Lakes. The structure rests on a concrete pier measuring 54 feet by 60 feet by 23 feet high. The lighthouse is a steel framed building encased in 1/4 inch steel plate. The lower segment measures 29 feet by 44 feet by 22 feet high, while the

tower supporting the lantern measures 14 foot square and extends an additional 20 feet above the first section. The round cast iron lantern has helical bars on the glass panels. It contains a Fourth Order Fresnel lens probably taken from the Milwaukee Pierhead Light. Overall, the building is 53 feet high from the base to the top of the ventilator ball and produces a lens focal plane 61 feet above the mean low water level of Lake Michigan.

North Point Light (1888, 1913)

The first lighthouse at North Point was constructed in 1855 to mark the entrance into the Milwaukee River at Milwaukee. The original brick structure was replaced and moved about 100 feet inland in 1888 when the station became threatened by the increased erosion of the beach. A 10-sided cast iron structure was built, measuring 14 feet in diameter at the base, 9 feet, 6 inches in diameter below the watchroom, and 39 feet high from the base to the top of the ventilator ball. This structure was adequate and remained in good condition, but by the early part of the twentieth century, the trees growing in the Lake Park area along the shore began to block the light, obscuring it from mariners on Lake Michigan. On March 4, 1909, Congress appropriated $10,000 to raise the light. This was done by erecting a new steel superstructure next to the old lighthouse and then utilizing the old structure as the upper portion of the taller lighthouse. Work began in July 1912 and was completed in April 1913 at a cost of $9,455, but the work was sufficiently complete to allow the light to be placed back into service on December 15, 1912. The new superstructure also has a 10-sided configuration and measures 21 feet, 6 inches in diameter at the base, 14 feet in diameter at the top, and is 35 feet high. The total height of the new structure was 74 feet from the base to the top of the ventilator ball, producing a focal plane 154 feet above the mean low water level of the lake, giving this light a range of about 21 miles in clear weather. The lens

Port Washington Breakwater Light

SHEBOYGAN BREAKWATER LIGHT

MANITOWOC BREAKWATER LIGHT

which is still in place was probably installed in 1888 and is a Fourth Order Fresnel, manufactured by Barbier, Benard & Turenne of Paris. It consists of four 90 degree panels, including two flash panels. The equipment still intact, although no longer used, includes a clockwork turning mechanism driven by a weight which fell down a central weight shaft, turning the lens and thus imparting the flash. Martin Knudsen ended his long career with the Lighthouse Service by serving as chief keeper at North Point from 1917 to 1924.

Port Washington Breakwater Light (1935)

Port Washington received its first lighthouse in 1850, and the first pier light was built in 1889, but this light was constructed after two new breakwaters were finished in 1934. It is a white square steel tower resting on a four-legged square concrete base and has a lens lantern with a focal plane 78 feet above the lake level. This station is typical of the modern design, and it is identical to the Indiana Harbor East Breakwater Light.

Sheboygan Breakwater Light (1915)

The first lighthouse at Sheboygan Harbor was built in 1839, but after the light had been rebuilt and moved several times, this structure and a skeletal steel pierhead tower were both constructed in 1915. This light was

140

originally a conical steel tower, with a diameter of 12 feet, 6 inches at the base. It was fitted with a round cast iron lantern, and had an overall height of 51 feet from the base to the top of the ventilator ball. Today, the tower is fitted with a glass lens lantern.

Manitowoc Breakwater Light (1918)

The Federal Government first built a lighthouse at the mouth of the Manitowoc River, in the Territory of Wisconsin, in 1840, one of the early Lake Michigan light stations. The first Manitowoc Breakwater Light was established in 1895, but the present structure was built after Congress authorized expenditures of $21,000 on June 12, 1917 for the construction of a new steel fog signal building and lighthouse. The building, completed in 1918, rests on a concrete pier measuring 22 feet by 48 feet by 11 feet high. The pier, in turn, rests on the concrete breakwall. The fog signal house is a steel-framed building encased in steel plates and measures 19 feet by 34 feet on the first floor, an area which served as the compressor room. The watchroom, which comprises the second floor of the structure, is 17 feet square and is surmounted by the round light tower which is 12 feet in diameter and supports the 10-sided lantern. Overall, the structure is 40 feet high, measured from the top of the pier to the top of the ventilator ball, and creates a lens focal plane 51 feet above the mean low water level of Lake Michigan. The lens, installed in 1918, was a fixed Fourth Order Fresnel, which was equipped with a 90 degree reflector. This lens was later removed, because the lens now in place is a Fifth Order Fresnel made by Sautter, Lemonnier & Co. of Paris.

Two Rivers Light (1883)

This light was erected at the end of the north pier at Two Rivers Harbor in 1883. It exhibited a fixed red

RAWLEY POINT (TWIN RIVER POINT) LIGHT

beacon, some 36 feet above the lake, that was visible for eight miles. It was replaced by a new steel tower built in 1969, and the old station was given to the community.

Rawley Point (Twin River Point) Light (1874, 1894)

The first lighthouse in this area was erected in 1853 "at the mouth of the Twin River, in the state of Wisconsin." Only the top ten feet of this square wooden tower has survived and the remains are now on display in Two Rivers, Wisconsin. The more significant structure to be built here was the coastal light constructed at Twin River Point (later renamed Rawley Point) in 1873-74, at a cost of $40,000. The original brick light tower has not survived, but the 1874 keeper's residence still remains at the site. It is a rectangular building with gabled and hipped roofs, measuring 39 feet by 65 feet overall. The present light tower was built in 1894 and was a reconstruction and enlargement of a lighthouse removed from the Chicago River entrance at Chicago. It is an octagonal pyramidal iron skeletal frame surmounted by a round cast iron watchroom and a 10-sided cast iron lantern. The watchroom may be reached by a set of circular stairs that are supported by a cast iron stair cylinder six feet in diameter. The eight corner posts of the skeletal framework rest on concrete piers, each eight feet square, and a distance of 38 feet apart from center to center across the diameter of the base. The corner posts are tied together with a complex system of steel horizontal struts and diagonal tie rods. The structure has two watchrooms. The first sits on top of the skeletal tower and measures 12 feet in diameter and 8 feet high. Above it is a second watchroom 9 feet, 6 inches in diameter. This watchroom, along with a lantern, was part of the lighthouse transplanted from Chicago in 1894. Overall, the tower stands 111 feet high from the base to the top of the ventilator ball and creates a focal plane 113 feet above the water level of the lake. This lighthouse was originally equipped with a Third

WATCHROOM AND LANTERN AT RAWLEY POINT

Order Fresnel Lens, but now exhibits a 36-inch airport-type beacon with a maximum range of 25 miles.

Kewaunee Pierhead Light (1909, 1931)

The building which now houses both the fog signal and light at Kewaunee Pier was constructed in 1909 and originally contained only the fog signal equipment. At that time there were a pair of range lights on the pier. The front light (1891) stood 24 feet and the rear light was 42 feet in height. The old front light was removed in 1931 and a new square tower was built on top of the east end of the fog signal building. The surviving station, measuring 10 feet by 54 feet and 23 feet high to the top of the roofline, consists of a lower half which is steel-framed and covered with one-quarter inch steel plate and an upper portion which is wood-framed with a shingle exterior. The steel tower, measuring 10 feet, 2 inches square, supports an octagonal cast iron lantern,

ALGOMA PIERHEAD FRONT LIGHT

producing a lens focal plane 45 feet above the level of Lake Michigan. The lens, perhaps taken from the 1891 lighthouse, is a fixed Fifth Order Fresnel that was made by Barbier and Fenestre of Paris.

Algoma Pierhead Front Light (1908, 1932)

The Algoma Pierhead Lighthouse, first established in 1893, was entirely rebuilt in 1908, and received substantial modifications in 1932. As rebuilt in 1908, it was a conical tower built of 5/16 inch steel plate, 8 feet in diameter at the base and 7 feet in diameter at the parapet, and stood 26 feet high from the base to the top of the ventilator ball. It supported a round watchroom surmounted by a 10-sided lantern housing a Fresnel lens. In 1932, the entire structure was raised to a height of 42 feet by placing the older tower on a new steel base 12 feet in diameter. The original lens has been replaced by a modern plastic lens.

Sturgeon Bay Canal North Pierhead Light (1882, 1903)

When the Sturgeon Bay-Lake Michigan Ship Canal opened in 1881, vessels passing from Green Bay into the open waters of Lake Michigan were given an alternate route so that they were able to avoid using the often-dangerous passage known as "Death's Door" located at the tip of the Door Peninsula. This lighthouse marked the entrance into the canal and was built in 1882, but it underwent major renovations in 1903. The original fog signal building measured 24 feet square and was surmounted by a square light tower which was 29 feet high. As a result of modifications made in 1903, the present fog signal building is 24 feet by 33 feet, with a hipped roof, and is surmounted by a round light tower made of one-quarter inch steel plate. It stands 39 feet high from the base to the top of the ventilator ball. The 10-sided cast iron lantern houses a Sixth Order Fresnel Lens made by Henri Le Paute of Paris. The lens is fixed,

KEWAUNEE PIERHEAD LIGHT

and has a brass reflector of 180 degrees circumference, so that the light cannot be seen from land. Both the lantern and lens are probably the original equipment installed at the station in 1882.

Sturgeon Bay Canal Light (1899, 1903)

The Sturgeon Bay Canal Light was constructed in 1899 by John P. McGuire of Cleveland. This station was an experiment in lighthouse design which failed because the architects did not take into account the heavy winds of this area, and as a result the structure had to be modified substantially in 1903. The original design was a cylindrical iron tower 78 feet tall to the base of the watchroom, which was 12 feet in diameter and 7 feet, 6 inches high, surmounted by a round cast iron lantern. The tower was supported by eight triangular lattice buttresses, each 16 feet, 6 inches along the side attached to the tower and 6 feet, 6 inches long at the base. These

supports were anchored in a concrete foundation 29 feet square and 8 feet deep by several foundation bolts that measured 7 feet in length. In addition, large wire ropes extending from the base of the watchroom to the foundation were used to give the structure additional stability. This design, however, was not adequate for the continuous stress produced by the strong winds, and it had to be abandoned in 1903, when the structure was almost entirely rebuilt. The original light tower, which was 8 feet in diameter and supported the entire structure, was converted into a cylindrical stair tower which supported only the staircase leading to the watchroom. The original watchroom, lantern, and lens were retained, but these were placed on top of a new skeletal steel framework, and the concrete foundation was extended to support the larger buttresses. This increased the diameter of the foundation from 25 feet to 36 feet. Overall, the tower is 98 feet tall from the base to the top of the ventilator ball and creates a lens focal plane 107 feet above the mean low water level of Lake Michigan. The lens is a Third Order Fresnel, with eight panels, and the lens rotates to impart a flash. It was manufactured by Henri Le Paute of Paris.

Bailey's Harbor Range Lights (1870)

The Lighthouse Board built a pair of range lights at Bailey's Harbor in 1870 at a cost of $6,000. When the range lights went into operation, the older Bailey's Harbor Lighthouse (1853) built on the east side of the harbor was abandoned and it now lies in ruins. The range lights are virtually identical to the ones built at Presque Isle Harbor on Lake Huron the same year. The front light was placed on a simple octagonal wooden tower. It was 21 feet high and had a range of 11 miles. The rear light, positioned 950 feet behind the front light, was placed atop a square wooden tower on the keeper's house. Overall, this light tower measured 35 feet in height, and was visible for a distance of 13 miles.

Cana Island Light (1870, 1901)

Cana Island Light, built in 1869-70, is located about halfway between North Bay and Moonlight Bay, two natural harbors of refuge on the east side of the Door Peninsula. Early keepers at this light which helps mark both harbors included William Jackson (1870-71), Julius Warren (1872-75), William A. Sanderson (1875-91), Jesse T. Brown (1891-1913), Conrad Stram (1913-18), and John Hahn (1919-?). The station is positioned in an exposed location and during a particularly violent storm in October, 1880, the waves actually broke over the top of the keeper's house. The conical yellow brick light tower, which stands 88 feet high overall, was sheathed with steel plates in 1901 to prevent further deterioration of the brickwork, much like the modifications done at Big Sable Point on the other side of the lake. This tower, now painted white, is surmounted by a 10-sided lantern exhibiting a flashing white light, now entirely automated. A brick passageway connects the tower with the dwelling, a rectangular one-story yellow brick building with a gabled roof. An oil storage shed made of stone and a brick outhouse have also survived. Except in times of high water, the island can be reached by a gravel causeway built during the First World War.

Pilot Island Light (1851, 1858)

The Death's Door, *Porte Des Morts,* Passage from Lake Michigan into Green Bay was first marked with a lighthouse on Pilot Island in 1850, but the original station was remodeled in 1858. Because this area was often engulfed in fog, the station received a fog bell in 1862. Some of the keepers at Pilot Island were Victor Rohn (1872-76) and Martin Knudsen (1889-1896). The isolation of this station and the many shipwrecks which occurred here are discussed earlier in this book. The station, now automated, is still in service. It consists of a

147

BAILEY'S HARBOR REAR RANGE LIGHT

BAILEY'S HARBOR FRONT RANGE LIGHT

rectangular yellow brick residence with an attached square brick light tower some 40 feet tall, topped by a 10-sided cast iron lantern.

Plum Island Rear Range Light (1897)

Plum Island, located in the middle of Death's Door, received a pair of range lights in 1897. Martin Knudsen, previously serving at Pilot Island, became the chief keeper at Plum Island where he remained through 1899. Later keepers included Hans J. Hanson, Charles E. Young, Joseph Boshka, and Charles Boshka. The double keeper's dwelling is no longer standing and the original front range light tower was replaced in 1964 by a steel skeletal tower. The rear range light, however, is an iron skeletal pyramidal structure surmounted by a cast iron watchroom, eight feet in diameter, and surmounted by an octagonal lantern. Both can be reached through a round cast iron stair tower. Overall, the light tower is 65 feet tall from the base to the top of the ventilator ball, creating a lens focal plane 80 feet above the water level. The lens is a fixed Fourth Order Fresnel made by Sautter, Lemonnier, and Co. of Paris.

Potawatomi (Rock Island) Light (1858)

In the early years of shipping in this area, the most popular route taken into Green Bay from Lake

CANA ISLAND LIGHT

Michigan was north of Washington Island. Potawatomi Island, named after the Indian tribe which lived in the area, lies just northwest of Washington Island. When this light went into service in the spring of 1837, the first keeper was David E. Corbin. After the station was eventually washed away, a new lighthouse was built in 1858. The gray limestone building remains, but the original lantern is gone, having been replaced by a lens lantern, still in use at this automated station. Although the light tower is only 41 feet tall, it creates a lens focal plane 159 feet above the lake surface.

St. Martin Island Light (1905)

The light tower on St. Martin Island is the only example of a pure exoskeletal tower on the Great Lakes. The six-sided tower is supported by six exterior steel posts which have latticed buttresses that rest on a concrete foundation 17 feet, 6 inches in diameter and 8 feet deep. The tower extends 57 feet to the base of the watchroom, which is 8 feet high and is surmounted by the round cast iron lantern with helical bars across the glass panels. Overall, the structure measures 77 feet from the base to the top of the ventilator ball and produces a lens focal plane 84 feet above the mean low water level. The lens is a Fourth Order Fresnel manufactured by Barbier, Benard & Turenne of Paris. It consists of two glass panels separated by brass reflectors, so that when turned, the lens produces alternating red and white flashes. It's interesting to note that while the white light may be seen for 24 miles, the red is visible for only 19 miles. The original manual weighted clockwork turning mechanism is in place, but it is no longer used. The station also includes a two-story rectangular brick keeper's dwelling measuring 32 feet by 52 feet, and a rectangular brick fog signal building that is 22 feet by 40 feet. Both have a gabled roof.

150

AERIAL VIEW OF PILOT ISLAND

POTAWATOMI (ROCK ISLAND) LIGHT

PLUM ISLAND REAR RANGE LIGHT

Poverty Island Light (1875)

This light was built in 1874-75 at a cost of $21,000 to mark the Poverty Island Passage into Green Bay. A conical brick light tower 70 feet tall is connected by a brick passageway to a one-story brick keeper's dwelling with a gabled roof. The original lantern was replaced by a lens lantern many years ago. In 1976, the light was placed atop a new skeletal steel tower built nearby.

Eagle Bluff Light (1868)

This light, which marks the east side of the Strawberry Channel, the east passage into Green Bay, was built in 1868 at a cost of $12,000. Early keepers at Eagle Bluff were Henry Stanley (1868-83) and William Duclon (1883-1918). The station includes a square yellow brick light tower 43 feet tall, oriented diagonally to, and built into the northeast corner of the keeper's house. The

151

ST. MARTIN ISLAND LIGHT

POVERTY ISLAND LIGHT

CHAMBERS ISLAND LIGHT

152

dwelling is a rectangular building, also of yellow brick construction, with a gabled roof. The original light, some 76 feet above the lake, was fixed and its beacon was visible for 16 miles. When the station was eventually automated by the Lighthouse Board in 1909, it was given a flashing white light, but its range was reduced to a distance of seven miles. This light remains in service. In 1961-63, the keeper's residence was restored to its original condition and is now a museum of Door County maritime and lighthouse history.

Chambers Island Light (1868)

In July, 1866, Congress approved funds for "a third-class lighthouse on *Mah-no-mah* or Chambers Island," to mark the west passage into Green Bay. Construction was completed in 1868. The keepers have included Lewis Williams, Peter Knudson, and Charles A. Young, all prior to 1895, Soren Christianson (1895-1900), Joseph Napeizinski (1900-06), and Jens J. Rollefson (1906-10). The octagonal brick light tower, with its lantern removed, has survived, along with the attached brick keeper's house. This lighthouse was originally equipped with a fixed white light, varied by a white flash every 60 seconds. It had a focal plane 68 feet above the water level and was visible for 16 miles. The Coast Guard placed a new light on a steel tower erected near the old lighthouse in 1961, when the older facility was abandoned. The new light tower exhibits a flashing white light some 97 feet above the water. The range of this beacon, however, is only 12 miles.

Sherwood Point Light (1883)

Congress appropriated $12,000 on March 3, 1881 for the construction of a lighthouse at Sherwood Point to mark the entrance into Sturgeon Bay from Green Bay. The station includes the rectangular brick keeper's dwelling with a gabled roof, measuring 25 feet by 37 feet in plan, plus the attached brick light tower supporting the cast iron 10-sided lantern. The tower produces a lens focal plane 61 feet above the water level. The lens is a Fourth Order Fresnel bearing the markings, "Barbier & Fenestre, Constr. Paris, 1880." The site also includes a frame bell tower. When the station was automated by the Coast Guard in the fall of 1983, it was the last manned light on the Great Lakes.

Green Bay Entrance Light (1935)

Much older aids have marked the entrance to Green Bay, including the Tail Point Lighthouse (1848) and the Grassy Island Range Lights (1872). This modern structure, which rests on a concrete pier, is a conical steel tower, exhibiting a lens lantern with a focal plane 72 feet above the water level.

153

SHERWOOD POINT LIGHT

PESHTIGO REEF LIGHT

Peshtigo Reef Light (1934)

This light station is located some two miles out into Green Bay off Peshtigo Point. It is a modern, white conical steel tower which rests on a round concrete crib. The tower stands 72 feet tall and exhibits a lens lantern equipped with a white light that flashes every six seconds. Its beacon is visible for ten miles. This station is also equipped with a fog signal.

Menominee North Pier Light (1927)

This lighthouse was first established in 1877, but the octagonal steel tower standing today was built in 1927. It is 25 feet tall, 15 feet wide, and rests on a rectangular concrete base, which is situated on a round concrete crib some 40 feet in diameter. The 10-sided lantern holds a Fourth Order Fresnel lens.

Minneapolis Shoal Light (1934)

The Minneapolis Shoal Light is typical of the design used by the Coast Guard for the last generation of manned light stations built on isolated reefs throughout the Great Lakes. It is identical to the station built at Gray's Reef in 1932, and is similar to many other stations as well. The steel-framed structure is encased in riveted steel plates and rests on a concrete pier 64 feet square and 12 feet high. The pier sits on a submerged crib which serves as the overall base of the structure. The first floor segment of the light station is 30 feet square, 15 feet high and is surmounted by the light tower which is 10 feet square and extends 41 feet to the base of the round lantern, which has helical bars and measures 7 feet, 11 inches in diameter. The resulting lens focal plane is 82 feet above the mean low water level. The lens is a Fourth Order Fresnel bearing the inscription "Sautter & Cie, Constructeurs a Paris."

MINNEAPOLIS SHOAL LIGHT

PENINSULA POINT LIGHT

GREEN BAY ENTRANCE LIGHT

Peninsula Point Light (1866)

This lighthouse, completed in 1866, consisted of a one-story keeper's residence and an attached square brick light tower, measuring 40 feet tall. The dwelling was totally destroyed by fire in 1959, so only the tower remains standing. The original oil lamp used here was replaced by an automatic acetylene lamp in 1922, but the station was abandoned in 1936 after Minneapolis Shoal Light went into service. The tower has a cast iron spiral staircase leading to the 10-sided cast iron lantern. This light is located in a picnic area which is part of the Hiawatha National Forest. The tower is open to visitors.

Manistique East Breakwater Light (1917)

As early as 1892, the Lighthouse Board recommended that a light station be built at Manistique, since there was no light in the northern part of Lake Michigan between Seul Choix Point and Poverty Island, a distance of approximately forty-four statute miles. The light was finally built in 1915-17 at a cost of nearly $20,000. It is a red steel tower, 14 feet square at the base and 10 feet square at the parapet, and it stands 35 feet from the base to the lantern deck. This station rests on a concrete crib and overall, produces a lens focal plane 50 feet above the level of Lake Michigan. It was originally equipped with a Fourth Order Fresnel lens. □

Menominee North Pier Light

8 THE LIGHTS OF LAKE SUPERIOR

Lake Superior was the second of the Great Lakes to be discovered by Europeans, after Lake Huron, but before Lakes Michigan, Erie, and Ontario. The first white men to explore Lake Superior were the French explorers Etienne Brulé and a man named Grenoble, who happened to arrrive at the western end of the lake first, in 1620, where they made contact with a tribe of Indians who were mining copper. By the 1650s, the remaining boundaries of the lake were well delineated by the French, and they had given it the name, "Lake Superior," meaning "the upper lake." Most of the early explorers and settlers were Jesuit missionaries intent on converting more than a dozen Indian tribes in the area, including the Huron, Chippewa, and Dakota (Sioux) to Christianity. Fur trading, missionary work, and warfare were the most important activities in the region during the sixteenth and seventeenth centuries. The commander of the French outpost at Chequamegon Bay, Wisconsin, launched an unsuccessful effort to mine copper on the Ontonagon River in 1737-38. Fighting between the French, British, and the major Indian tribes for control over the area temporarily abated after the British acquired French Canada in 1763. Alexander Henry explored for copper on the Ontonagon River in 1766 and a party of miners worked there in 1771-72 without success. Their failure, combined with the disruptive influences of the American Revolution and the War of 1812, discouraged additional efforts.

Lewis Cass, the Michigan Territorial Governor, and Henry R. Schoolcraft systematically explored the south shore of Lake Superior in 1820. After Michigan achieved statehood in 1837 and acquired the western Upper Peninsula in the process, Governor Stevens T. Mason appointed Douglass Houghton as Michigan's first State Geologist. Houghton, who had served on a second Schoolcraft expedition to Lake Superior in 1831, began surveying the mineral resources of the Upper Peninsula in the summer of 1840 and issued a report in 1841 suggesting extensive copper deposits on the Keweenaw Peninsula. After the Chippewa gave up their rights to their lands west of the Chocolay River to the United States Government in the Treaty of La Pointe in March, 1843, the last obstacle to exploration was gone. After an initial "copper rush" involving hundreds of speculators in 1843-45, mining companies gradually discovered and developed the copper deposits in the 1850s and 1860s.

Meantime, the Jackson Iron Company of Jackson, Michigan, discovered iron deposits a few miles west of Marquette in June, 1845, and quickly opened a mine near present-day Negaunee. Ore production from the Marquette Range was marginal until the canal at Sault Ste. Marie opened in April, 1855. Shipments of iron ore then jumped from about 1,500 tons in 1855 to 236,000 tons in 1865. Additional iron ore deposits had been found in the vicinity of Iron Mountain and Iron River, both on the Menominee Range, but these sites were not developed until 1877, when railroad connections were completed to Escanaba on Lake Michigan. The deposits of the Gogebic Range, on the Michigan and Wisconsin sides of the Montreal River, were first developed in 1884 and this became a major producing region, with most of the ore shipped by rail to Ashland, Wisconsin on Lake Superior, and from there by water to the rest of the Great Lakes system. By the 1880s, Michigan was supplying about 40 percent of American iron ore output and half of the nation's copper.

The iron deposits found in the Vermilion Range in Minnesota were discovered in the early 1880s, but were dwarfed by the subsequent discoveries in the Mesabi Range in the early 1890s. The volume of iron ore produced near Lake Superior and shipped over its surface grew enormously, as evidenced by the numbers of vessels plying the lake. The canal at Sault Ste. Marie recorded 193 ship passages in 1855, about 5,400 in 1885, and nearly 18,000 a decade later. The number of ship passages peaked at 25,407 in 1916, but the tonnage

Marquette Harbor Light

continued to climb through the 1920s. The iron mines of the Lake Superior region produced between one-quarter and one-third of American iron ore output through the Second World War. The Gogebic and Menominee ranges stopped production in the late 1960s, and production from the Marquette and Mesabi ranges has fallen, but they still generate considerable lake traffic.

The treacherous St. Mary's River, the connecting waterway between Lake Superior and Lake Huron, is a meandering route filled with sharp turns and many islands, so it has long required many lights and buoys. By 1898, for example, the river had 24 pair of range lights and an additional 12 single lights, for a total of 60 aids to navigation. The Coast Guard *Light List for 1978* identifies 88 lights along this route. The vast majority of these are steel skeletal towers, steel or concrete columns, and wooden posts, mostly of recent construction. They generally stand less than twenty-five feet tall and have relatively short ranges.

Round Island Light (1892)

This structure is the only light station on the St. Mary's River that will be discussed in this chapter. It was built on the northwest side of Round Island at the lower end of the river in 1892. The wood-framed keeper's dwelling has a square tower built into the long side of the rectangular building, near the water. It still has its original octagonal lantern, but the lens is no longer in place. Instead, a skeletal steel tower built nearby in 1923 exhibits a lens lantern.

Point Iroquois Light (1871, 1902)

The Ojibwa Indians massacred a group of invading Iroquois warriors at this location in 1662, giving the point its original Indian name. It was noted by the early explorers, but the location did not become significant until the St. Mary's Falls Ship Canal opened in 1855.

160

ABANDONED REAR LIGHT AT CEDAR POINT (ROUND ISLAND POINT)

POINT IROQUOIS LIGHT

Vessels passing from Lake Superior into the canal came close to Point Iroquois, through a narrow passage lined with sandy shores on the American side and rocky reefs on the Canadian side. The original Congressional appropriation of March 3, 1853, provided $5,000 to build a light either at Point Iroquois or on an unspecified island in Lake Michigan off Point Aux Chenes, near the Straits of Mackinac. The Lighthouse Board decided that the Lake Superior light was more urgently needed and the construction of a simple wooden house with an attached tower was completed in 1855. The original facility was entirely replaced by a brick tower and attached residence constructed in 1871 at a cost of $18,000. The conical tower, 16 feet in diameter at the base, stands 51 feet tall to the lantern deck and 65 feet high overall. In 1902, a two-story rectangular brick house was added to the residence to provide housing for the assistant keeper. The Coast Guard gave up this facility in 1965 and transferred the property to the United States Forest Service.

Whitefish Point Light (1861)

The area around Whitefish Point has long been known as the Graveyard of the Lakes because of the numerous dangerous shoals in the area. Since the beginnings of regular navigation on Lake Superior, seventy major shipwrecks have occurred near here,

including the loss of the *Edmund Fitzgerald* on November 10, 1975. Congress appropriated $5,000 for a light in March, 1847, and Ebenezer Warner of Sandusky, Ohio, signed a contract with the Federal Government on August 21, 1847, to build a lighthouse ''of split stone or hard bricks and laid in good lime mortar.'' Before completing the job, Warner received an additional sum of $3,298 for his work. Cost overruns were not entirely unknown in the nineteenth century. The lighthouse was completed in late 1848 and placed into service early in 1849, at the same time as the light at Copper Harbor, so each can lay claim to being the first light on Lake Superior. James B. Van Ranselaer was appointed keeper at Whitefish Point in October, 1848 and served until May, 1851. The tower was replaced in 1861 by an ''iron pile'' tower better suited to withstand Lake Superior's severe winter weather. Identical towers were also built in 1861 at DeTour Point and on Manitou Island, and the latter tower has survived as well. The tower has a pyramidal skeletal frame, with horizontal and diagonal bracing, supporting the cast iron watch room and lantern, which are reached through a round cast iron stair tower standing forty-two feet tall. Whitefish Point originally had Winslow Lewis lamps and reflectors, before receiving a Fresnel lens in 1857, and in 1913, the Lighthouse Service installed a 1,000 watt Aladdin incandescent oil vapor lamp. The station has a two-story frame keeper's residence and was

WHITEFISH POINT LIGHT

CRISP'S POINT LIGHT

162

automated by the Coast Guard in 1970. The Great Lakes Shipwreck Historical Society now operates a museum at the station.

Crisp's Point Light (1904)

In 1896 the Lighthouse Board recommended that a light be built at Crisp's Point, about thirteen miles west of Whitefish Point, because many vessels heading for Whitefish Point had deviated slightly from their course and were wrecked at Crisp's Point. In June, 1902, Congress appropriated $18,000 for this purpose and construction began in 1903. The new light went into service on March 5, 1904, exhibiting a fixed red Fourth Order Fresnel lens made by Sautter & Lemonnier of Paris. This lens has since been replaced with a 300 millimeter plastic lens. When built, the station included a keeper's house, fog signal building, and a variety of other outbuildings, but these have been demolished and all that remains is the light tower and a small attached rectangular brick building. The conical brick tower is 58 feet high from the base to the top of the ventilator ball and produces a focal plane 58 feet above the mean low water level of Lake Superior. It is 14 feet in diameter at the base and 9 feet in diameter at the parapet, rests on a concrete foundation 10 feet deep, and is surmounted by an octagonal cast iron lantern with an inscribed diameter of 7 feet. At the base of the tower, the outer wall is 18 inches thick, with an air space of 22 inches, and an inner wall, 8 inches thick, supports the staircase. At the parapet, the outer wall narrows to a thickness of 8 inches, the air space is 6 inches, and the inner wall is only 4 inches thick.

Grand Marais Harbor Range Lights (1895, 1898)

Grand Marais is the only harbor of any substantial depth between Whitefish Bay and Munising. This pair of range lights are both white, square steel pyramidal

skeletal towers with the upper portion enclosed in steel plates. The outer light, located on the end of the pier, was built in 1895 and stands 34 feet tall overall, while the inner light, built in 1898, is 47 feet high.

Au Sable Point Light (1874, 1909)

In the 1860s and 1870s there were no lighthouses along the heavily used coast between Whitefish Point and Grand Island. In June, 1872, Congress agreed to spend a total of $40,000 for a light at Au Sable Point. Construction began in July, 1873, and the structure was finished in August, 1874. The conical brick tower at this station is 16 feet in diameter at the base and stands 87 feet tall, creating a lens focal plane 107 feet above Lake Superior. The attached brick lightkeeper's house was originally a single-story building, but was raised to two stories in 1909, when a second keeper's house was added to the station. The complex includes a fog signal

Grand Marais Harbor Inner Range Light

building and assorted outbuildings. Civilian keepers included Casper Kuhn (1874-76), Napoleon Beedon (1876-79), Frederick W. Boesler, Sr. (1879-83), Gus Gigandet (1884-96), Herbert W. Weeks (1897-1903), Otto Bufe (1903-05), Thomas E. Irvine (1905-10), James Kay (1910-15), John Brooks (1915-23), Klass Hamringa (1923-30), Arthur Taylor (1930-36), and Edward T. McGregor (1937-45). The Coast Guard automated the station in 1958 and transferred the property to the Pictured Rocks National Lakeshore.

Grand Island North Light (1868)

Grand Island, just off the south shore of Lake Superior, was a favorite stopping point for the earliest French explorers. Abraham Warren Williams from Vermont settled on the island in 1840, and established a fur trading post. His family owned the island until 1900, when it was sold to the Cleveland-Cliffs Iron Company.

The first light was built on the north end of the island in 1855 at a cost of $5,000. Although the original wooden tower was only forty feet tall, it produced a focal plane 204 feet above Lake Superior because the tower rested on the tall cliffs. It is still one of the highest lights on the Great Lakes. The surviving rectangular brick keeper's house with attached square brick light tower was constructed in 1867-68 at a cost of $17,000. Reuben Smith was appointed keeper in 1867, Frederick Ballard held the post from 1885 to 1893, and George Gentry was the keeper around 1905. The lamps, which had used kerosene since the 1870s, were converted to acetylene in 1920. The Coast Guard automated the facility in 1961, and it remains in use today.

Grand Island East Channel Light (1868)

In June, 1860 and July, 1866, Congress appropriated a total of $16,000, ''for one or two beacon lights, at the discretion of the Secretary of the Treasury, at the entrance to Grand Island Bay and Harbor, Lake Superior.'' A light or set of lights was needed to allow vessels to use Munising Harbor as a major harbor of refuge. The Lighthouse Board spent the funds in 1868-70 building the East Channel Light and a pair of range lights on shore. The lighthouse, a wood-framed keeper's house with an attached square wooden tower, operated only until 1913, when it was replaced by the new range lights installed at Munising in 1908. George Pryor was working as keeper at the East Channel Light at the turn of the century and was probably the last person to serve at the post.

Munising Range Lights (1908)

The range lights at Munising were built to guide ships into the harbor through a narrow channel running east of Grand Island. Congress appropriated $15,000 for this

165

MUNISING FRONT RANGE LIGHT

TOWER INTERIOR AT MUNISING FRONT RANGE LIGHT

purpose in March, 1907, and the contract for the two towers was awarded to Champion Iron Company of Cleveland, Ohio, with all work to be completed in August, 1908. The front range light is housed in a conical tower made of 5/16 inch riveted steel plates and measures 12 feet in diameter at the base and 8 feet in diameter at the parapet. It is surmounted by an octagonal cast iron lantern with an inscribed diameter of 7 feet. Overall, the tower is 58 feet high from the base to the top of the ventilator ball and creates a lens focal plane 79 feet above the mean low water level of Lake Superior. When built, both the front and rear range lights were equipped with Adam and Westlake 23-inch reflectors, but both now have locomotive style headlamps with the inscription, ''Golden Glow, Essco, Phila.'' Next to the front range light there is a keeper's dwelling, a two-story rectangular brick and frame building with a gabled roof, measuring approximately 20 feet by 35 feet. The rear range tower is also a conical

structure of riveted steel plates 5/16 inch thick, with a diameter of 10 feet 9 inches at the base and 8 feet at the base of the lantern, which is 7 feet in diameter with a single window (34 inches by 42 inches) serving as the only lantern panel. It stands 33 feet tall, but because it is located uphill from the front range light, it has a higher focal plane of 107 feet above the level of the lake.

Marquette Breakwater Outer Light (1908)

There has been a light tower to mark the end of the Marquette Breakwater since 1875, but the structure standing today was erected in 1908. It is a square pyramidal skeletal tower, with the upper portion enclosed with steel plates to form a watchroom. The watchroom is 8 feet high, 14 feet in diameter at the base, and 10 feet in diameter at the top, and it is surmounted by an octagonal cast iron lantern. Overall the structure is 36 feet high and produces a lens focal plane of 40 feet

Munising Rear Range Light

above the mean low water level. The Fourth Order Fresnel lens which is exhibited was manufactured by Henri Le Paute of Paris and consists of three panels creating an arc of illumination of some 270 degrees to prevent the light from being seen in Marquette.

Marquette Harbor Light (1866, 1906)

Marquette quickly became an important harbor and commercial center in the early 1850s with the discovery of the Marquette Range iron ore deposits and the opening of the St. Mary's Falls Ship Canal at Sault Ste. Marie in 1855. The first lighthouse at Marquette was erected in 1853, but was already in poor condition when Congress appropriated $13,000 for its reconstruction in April, 1866. The light tower standing today is the 1866 reconstruction, while the keeper's house was changed considerably in 1906 when a second story was added. The brick light tower is 9 feet, 4 inches square and 38 feet, 9 inches high, with walls 13 inches thick at the base. It is surmounted by a 10-sided cast iron lantern with an inscribed diameter of 7 feet, and produces a lens focal plane of 70 feet above Lake Superior. At the time of the 1906 reconstruction, this lighthouse was equipped with a Fourth Order Fresnel lens driven by a clockwork turning mechanism made by Barbier, Benard & Turenne of Paris. An incandescent oil vapor lamp was installed in

1907, but it is now fitted with a Westinghouse 36-inch Airway Beacon rated at 703,000 candlepower.

Presque Isle Harbor Breakwater Light (1941)

The first breakwater in Presque Isle Harbor, just north of Marquette, dates from 1896, when the Lake Superior and Ishpeming Railroad built its first iron ore dock at Presque Isle Harbor. The present breakwater was constructed in 1926, and stood about 1,000 feet long, but it was extended to 2,600 feet in 1935. The light, built to mark the breakwater, was finished in 1941 and is a steel cylindrical tower atop an octagonal building, producing a lens focal plane 56 feet above Lake Superior.

Granite Island Light (1868)

In their *Report For 1866,* the Lighthouse Board noted the need for a light to mark this very small island located twenty miles northeast of Marquette, and dangerously near the coastal shipping lanes. Congress responded by appropriating $20,000 for this purpose on March 2, 1867 and construction was begun. One of the major problems encountered was the need to blast away the top of the island to create a reasonably flat construction site. Work was completed in the spring of 1868 in time for the

168

PRESQUE ISLE HARBOR BREAKWATER LIGHT

GRANITE ISLAND LIGHT

STANNARD ROCK LIGHT

HURON ISLAND LIGHT

opening of the navigation season on Lake Superior. The light tower and keeper's house were both built of rough coursed granite blocks, with coursed ashlar used on the cornices and also to frame all of the windows and doorways. The light tower is 10 feet square and 40 feet high, with walls 21 inches thick at the base, surmounted by a 10-sided cast iron lantern, producing a lens focal plane of 89 feet above the mean low water level of Lake Superior. The original lens was a Fourth Order Fresnel, with flash panels and a clockwork turning mechanism, manufactured by Le Paute of Paris. There is now a plastic lens lantern in service, placed on top of the lantern. The attached keeper's house is a rectangular building with a gabled roof, measuring 28 feet by 43 feet. This light began to lose its significance in the 1920s when shipping ran less often between the island and the shore, but instead passed well to the east of the light station. The Coast Guard automated the facility in 1939 and the island reverted to the gulls.

Stannard Rock Light (1882)

On August 26, 1835, Charles C. Stannard, Captain of the American Fur Company's schooner *John Jacob Astor*, first spotted a dangerous reef, more than a mile in length, which rose slightly above the water level at its northern end. The rock at the end of the reef received his name. Located 23 miles southeast of Manitou Island and 55 miles north of Marquette, along a major shipping route, the reef was immediately recognized as a threat to navigation. In its report for 1866, the Lighthouse Board described it as "the most serious danger to navigation in Lake Superior," and in 1868 placed a day beacon there. Detailed surveys of the reef were conducted in 1874 under the supervision of Major General Godfrey Weitzel of the U.S. Army Corps of Engineers at a cost of $10,000 and Congress appropriated $50,000 in 1877 to begin construction of a lighthouse. The project used the equipment developed for the Spectacle Reef Light

169

MENDOTA (BETE GRISE) LIGHT

completed in 1872, but the Board recognized that this one would be more difficult and costly. Their $300,000 estimate of total costs in 1874 was pretty close to the mark, however, for when this light was completed in 1882, the total bill came to $305,000.

Captain John A. Bailey, who had supervised the Spectacle Reef lighthouse construction, directed the work at Stannard Rock as well. He established a quarry on Huron Island to supply ballast stone and cement, using Skanee in Huron Bay as a base of operations. Significant progress did not occur until 1878, when the lighthouse tender *Warrington* towed an elaborate watertight cofferdam built at Skanee out to the site and sank it, using nearly 5,000 tons of rock as ballast. In the construction seasons of 1879 and 1880, a pier consisting of iron, stone, and 7,246 tons of cement was finished. The tower was made from enormous granite stones, each weighing up to thirty tons, which were cut, dressed, and pre-fitted at the Marblehead quarry on Lake Erie before being shipped to Lake Superior. The workmen laid all thirty-three courses of stone in July and August, 1881 and the remaining work was completed that fall and the following spring.

The light was first exhibited on July 4, 1882. It was a Third Order Fresnel lens made by Le Paute that rotated by means of a clockwork mechanism and was fueled with kerosene. Because of the extreme isolation and hazards of this station, few keepers have served here for

very long. Two exceptions were Elmer A. Sormunen (1934-54) and Louis Wilks (1936-53). The Stannard Rock Light was being automated on June 18, 1961, when a propane gas explosion occurred killing one of the four Coast Guardsmen stationed there and seriously injuring two others. Much of the equipment in the lower levels of the structure was destroyed and the light went out, but two days passed before the tragedy was reported. The station was fully automated in 1962 and now exhibits a 300 millimeter plastic lens.

Big Bay Point Light (1896)

The light at Big Bay Point was authorized by Congress in 1895, begun in May, 1896, and put into service in October of the same year. The keeper's dwelling is a two-story rectangular brick building, 22 feet by 52 feet overall, with 18 rooms. The square brick light tower supports a round steel watchroom and lantern, which originally had a Third Order Fresnel lens. A small brick fog signal building has also survived. The Coast Guard automated this station in 1941 and sold it in 1961, after building a new steel tower nearby.

Huron Island Light (1868)

The Huron Islands help mark the turning point for vessels travelling along the important coastal route

171

GULL ROCK LIGHT

MANITOU ISLAND LIGHT

PORTAGE LAKE LOWER ENTRANCE LIGHT

between Marquette and Keweenaw Bay. As traffic increased along these shores in the 1860s, the Lighthouse Board noted that the Huron Islands "are a constant source of anxiety to navigators, wrecks having frequently occurred at this point." The light was constructed on the highest granite outcropping of the west island and was lit for the first time on October 28, 1868. Built at a cost of $17,000, the station consisted of a keeper's house of grey granite blocks and an attached square light tower of the same material. Although the tower was only 39 feet tall, its location placed the lens 197 feet above Lake Superior. One tragedy occurred here during a blinding snowstorm on April 30, 1909, when the tow barge *George Nester* was wrecked just offshore. The Huron Island lightkeeper, Frank Wittie, and his assistant, Casper Kuhn, attempted to rescue the ill-fated crew of seven, but the *Nester* went down in less than five minutes. The original 20,000 candlepower kerosene lamp was replaced in 1961 by a 45,000

candlepower electric oscillator light. This station was manned by three keepers and later, by five Coast Guardsmen, until it was fully automated in 1972.

Sand Point Light (1878)

The light at Sand Point, built to mark the southern end of L'Anse Bay, was completed in 1878 at a cost of $10,000. One of the early keepers here, J.B. Crebassa, served from 1879 through 1899. The station consists of a small red brick keeper's house with an attached square brick tower at the eastern end of the dwelling.

Portage Lake Lower Entrance Light (1920)

Several lights have existed here to help guide ships passing into and through the Portage River entrance into Portage Lake. A pair of range lights went into service in this area in 1868 and were subsequently

PORTAGE RIVER (JACOBSVILLE) LIGHT

COPPER HARBOR LIGHT

STAIRS LEADING TO TOWER AT COPPER HARBOR LIGHT

replaced by another set. This structure, dating from 1920, is an octagonal steel tower standing 31 feet tall.

Portage River (Jacobsville) Light (1870)

In 1853, Congress initially approved construction of a lighthouse at the entrance to the Portage River, the water route connecting Lake Superior to the twin cities of Houghton and Hancock, in the heart of the Copper Country, and the work was done in 1855-56. The original station, made of white stone, was then replaced with the present building in 1870, at a cost of $12,000. Two of the early keepers were J.B. Crebassa (1873-78) and George Craig (1878-90). The surviving structures include a single-story rectangular brick keeper's dwelling, 25 feet by 40 feet, and an attached round brick tower standing 65 feet tall. The station now serves as a private residence.

Mendota (Bete Grise) Light (1895)

The first light in this area was a wooden tower built in 1870 to guide vessels into the Mendota Ship Canal linking Lac La Belle with Lake Superior. Operation of the light was discontinued in 1871, but was reestablished in 1895 with the installation of the surviving lighthouse, a T-shaped brick dwelling with a square yellow brick tower attached to its eastern end. The 10-sided cast iron lantern has survived, but with no lens. The Coast Guard abandoned this facility after new pierhead lights went into service at the entrance to Lac La Belle in 1960 and the old lighthouse is now in private hands.

Gull Rock Light (1867)

Gull Rock was a serious threat to ships attempting to pass between the tip of the Keweenaw Peninsula and Manitou Island in the course of going from the western end of Lake Superior to Keweenaw Bay. Congress

COPPER HARBOR REAR RANGE LIGHT

appropriated $15,000 in 1866 for the construction of a lighthouse on Gull Rock and work was carried out the following year, although delayed by the death of the construction foreman William Tunbridge. The light was first put into service on November 1, 1867 and exhibited a Fourth Order Fresnel lens manufactured by Barbier and Fenestre of Paris. This lens was eventually replaced by a 250 millimeter plastic lens. The light tower stands 46 feet in height, measured from the base to the top of the ventilator ball and creates a focal plane 50 feet above the mean low water level of Lake Superior. The tower is a square brick structure, with an outer wall 12 inches thick supporting the lantern and an inner wall 4 inches thick supporting the stairs. The cast iron lantern is 10-sided, with vertical bars. There is a two-story brick keeper's dwelling with a gabled roof attached to the light tower. The dwelling has suffered severe interior deterioration because it has been unoccupied and exposed to the ravages of the weather for many years.

Manitou Island Light (1861)

The first lighthouse on this island, located east of the tip of the Keweenaw Peninsula, was erected in 1850 at a cost of $7,500. The first keeper at this isolated station was Angus Mcleod Smith, who served from September, 1849 to October, 1856. Some of the other keepers in the early years included Elias Bouchard (1856-59), E.

Guilbault (1859-61), Henry Letcher (1861-64), Arnold Bennett (1864-66), and Charles Corgan (1866-76). The structure standing today was built in 1861 and along with the light tower at Whitefish Point built in the same year, is the oldest iron skeletal light tower on the Great Lakes. It consists of a square segment at the base, measuring 26 feet by 26 feet by 17 feet, six inches high, which in turn supports a four-legged pyramidal skeletal frame 42 feet, 6 inches high, surmounted by a cast iron watch room and cast iron lantern, both with ten sides. The four corner posts are braced with horizontal struts of cast iron pipe, each 12 inches in diameter at the first level and 4½ inches in diameter at the upper levels. The diagonal bracing consists of two inch wrought iron tie rods with turnbuckles. The watchroom is reached by a circular staircase enclosed by a stair cylinder of cast iron boiler plates one-quarter inch thick and lined with wood. The stair cylinder is 6 feet in diameter and 42 feet high. Overall, the tower is 80 feet high from the base to the top of the ventilator ball and produced a lens focal plane 81 feet above the mean low water level of Lake Superior. The original lens, removed in 1935, was a Third Order Fresnel made by Le Paute and had six separate panels, each with a bull's-eye prism in the central drum. The lens now in place is also a Third Order Fresnel, but has four panels and bears the inscription, "Barbier & Co., Paris." In addition, there is a two-story rectangular frame keeper's dwelling, probably built in 1861,

175

EAGLE HARBOR LIGHT

measuring approximately 30 feet by 50 feet, and linked to the base of the stair cylinder by a covered walkway. Some of the other keepers to see service here have included Henry Pearce (1876-79), George Howard (1879-81), Nathaniel Fadden (1884-86), Noah Bennett (1887-93), and Norman W. Smith (1893-1908).

Copper Harbor Light (1849, 1867)

This harbor became the focal point of early explorations for copper on the Keweenaw Peninsula beginning in June, 1843, when the Federal Government established a Mineral Land Agency office there. Congress appropriated $5,000 for a light at Copper Harbor on March 3, 1847. After competitive bidding, Charles Rude was awarded the contract for $4,800 and began work in August, 1848. The stone light tower which went into service in the spring of 1849 was located some 100 feet closer to the water's edge than the

surviving lighthouse. The station received a Fresnel lens in 1856. Remarkably, the 1849 detached keeper's house is still standing nearby. Early keepers here included Henry Clow (1849-53), Henry C. Shurter (1853-55), and Napoleon Beedon (1856-69). The Lighthouse Board replaced the 1849 structure with an entirely new building in 1866-67 at a cost of $13,700, probably because the water had undermined the foundations of the original tower. The surviving yellow brick keeper's dwelling and attached square light tower enjoyed a much better fate. It was well-maintained by keepers such as John Power (1868-76), Charles Corgan (1876-81), Edward Chambers (1881-82), and James W. Rich (1882-83). The light was discontinued from October, 1883 through June, 1888, but the property was kept up by the keeper of the Copper Harbor Range Lights. After this station reopened, Henry Corgan enjoyed a long tenure as keeper, serving from 1888 until his retirement in 1919, when the light was converted

STOREHOUSE AND FOG SIGNAL BUILDING AT EAGLE HARBOR

from kerosene to acetylene gas and left unattended. In 1927, the Lighthouse Service moved the gas light to a sixty foot steel tower nearby and leased the keeper's dwelling to a Chicago area physician who used it as a summer residence for two decades. The Michigan Department of Natural Resources acquired the property in 1963 and it is now used as a maritime museum that is part of Fort Wilkins State Park.

Copper Harbor Range Lights (1869, 1927, 1964)

Congress appropriated $3,500 in June, 1860 for a pair of range lights at Copper Harbor, but the actual construction was not carried out until 1865. Initially the War Department let the keeper use one of its buildings at Fort Wilkins as a residence, but in 1869, the Lighthouse Board built a keeper's dwelling with a tower on it to exhibit the rear range light. A skeletal tower erected in 1964 replaced the rear range light, and a smaller steel tower built in 1927 replaced the front range light structure. The first keeper at the range lights was Napoleon Beedon, who also ran the main light at Copper Harbor. The range light keepers were few in number, but long of tenure, and included William Tresize (1870-85), Charles Davis (1885-1930), and a man named Haven (1930-37), who was the last keeper. The 1869 keeper's dwelling, which held the rear range light, is still standing.

Eagle Harbor Light (1871, 1895)

The first lighthouse at Eagle Harbor, authorized by Congress in March, 1849, was completed in 1851 at a cost of $4,000. The original Winslow Lewis lamps and reflectors lasted only until 1857, when the Lighthouse Board installed Fresnel apparatus. By 1868, the Board reported to Congress that the original buildings were in poor condition and needed to be rebuilt. Congress appropriated $14,000 for this purpose in July, 1870 and the reconstruction was completed the following year. The octagonal brick light tower is ten feet in diameter, with walls 12 inches thick and it supports a 10-sided cast iron lantern. Overall, the tower measures 44 feet high from the base to the top of the ventilator ball and creates a lens focal plane 60 feet above the mean low water level of Lake Superior. The original lens, a Fourth Order Fresnel made by Sautter of Paris, has been replaced with a modern airport beacon lens. The keeper's dwelling is a rectangular brick building with a gabled roof covered with sheet tin. In 1890, the Lighthouse Board recommended the addition of a fog signal at Eagle Harbor, but actual construction was delayed and the new fog signal was not put into service until November, 1895. The fog signal building, a rectangular steel-framed building encased in corrugated sheet iron with a hipped roof, is still present, but none of the fog signal apparatus has survived. The complex also includes a

ROCK OF AGES LIGHT

brick oil house with a gabled roof, built in 1871. There are also two frame houses, each two stories tall, but these were built in the 1940s and moved to their present location at a later date.

Rock of Ages Light (1908)

The Lighthouse Board first recommended the construction of a lighthouse on the Rock of Ages in 1896, on the grounds that a light on this dangerous rock would enable ships in the western end of Lake Superior to run on the sheltered side of Isle Royale and away from the wind during severe weather. They first estimated that a light could be built for around $50,000, but this estimate proved to be hopelessly unrealistic. Congress finally appropriated $25,000 in March, 1905 to permit planning and design work to proceed and then followed with an appropriation of $100,000 in June, 1906 to cover construction costs. In March, 1909, an additional

appropriation of $15,000 was required to purchase a lens and the total cost for the project ultimately amounted to $138,329. The light tower was designed by Major Keller of the U.S. Army Corps of Engineers, while the construction was carried out by Walter F. Beyer of Detroit. The erection of this light was a major engineering feat largely because of the extreme isolation of the site from populated areas. The top of the rock had to be blasted away before the massive foundation pier could be built. It is a cylindrical structure measuring 50 feet in diameter and 30 feet in height, consisting of 3/8 inch steel plate filled with concrete. The conical tower, 30 feet in diameter at the base and 130 feet high, is made of steel plate encased and lined with brick below the watchroom. The round cast iron lantern has a diameter of 12 feet, 3 inches. The lens was a Second Order Fresnel made by Barbier, Benard & Turenne of Paris. It consisted of four panels, each with a bull's-eye lens in the center and rotates once every 20 seconds, producing

178

ISLE ROYALE (MENAGERIE ISLAND) LIGHT

ROCK HARBOR LIGHT

a double flash every 10 seconds. Each of the four panels of the lens has a center segment made up of a bull's-eye surrounded by six concentric prisms, with additional segments radiating out from the central segment and containing between 6 and 17 additional prisms. The combined effect of a powerful lens and a focal plane of 117 feet above the mean low water level of Lake Superior has produced a light with a range of 29 miles in clear weather, the most powerful beacon on the Great Lakes. Although the light tower was sufficiently completed to exhibit a temporary light in October, 1908, it was not until September, 1910 that the permanent lens was exhibited. The lens was removed by the National Park Service in June, 1985 and is on exhibit at the Windigo Ranger Station on Isle Royale. One interesting incident took place at Rock of Ages when the freighter *George M. Cox* ran aground on the nearby reef on May 27, 1933, and keeper John F. Soldenski helped rescue the 125 survivors, who spent a day packed together inside the light tower until taken off the next day.

Isle Royale (Menagerie Island) Light (1875)

The lighthouse on Menagerie Island marks the entrance into Siskiwit Bay, one of the major harbors of refuge on Isle Royale. Construction of this light station began in the spring of 1874, and the light was first exhibited on September 20, 1875, with the total cost of

the work amounting to $20,000. The two earliest keepers at this station were William Stevens (1875-78) and John H. Malone (1878-93). Malone's career and life at this station were discussed in an earlier chapter. The surviving light tower is an octagonal red sandstone structure 16 feet in diameter at the base, surmounted by a cast iron lantern of ten sides. The outer wall of the tower, bearing the load of the watchtower and lantern, is 40 inches thick at the base and 10 inches thick at the parapet, separated by a two inch air space from the inner wall, which supports the stairs. The keeper's dwelling is rectangular, with a hipped roof and is constructed of rough red sandstone. It is connected to the light tower by a covered passageway eight feet long. Overall, the tower is 61 feet high and produces a focal plane 72 feet above the lake. The lens is a Fourth Order Fresnel manufactured by Henri Le Paute of Paris. The Coast Guard automated this light station in 1941.

Rock Harbor Light (1855)

More than a dozen copper companies established mining operations on Isle Royale in the late 1840s and vessels entered Rock Harbor through the rock-infested Middle Islands Passage, mainly to supply the Smithwick and Ransom Mines, two of the earliest in the area. Congress spent $5,000 to build a lighthouse here in 1855, making the surviving tower and residence among

PASSAGE ISLAND LIGHT

TRAMWAY CART AT PASSAGE ISLAND LIGHT

the oldest still standing on the Great Lakes. Once the initial copper deposits were depleted and shipping consequently disappeared, the Lighthouse Board closed the station in 1859. After a second mining boom on Isle Royale developed in the early 1870s, the Board spent an additional $5,000 in 1874 to repair the light and place it back into service. They again abandoned the station in 1879, and it remained vacant for several decades, but in the early part of the twentieth century, summer camping parties used it as a shelter. Commercial fishermen Arnold and Milford Johnson then lived at the station from 1928 until 1939, when the property passed to the newly established Isle Royale National Park. The station has remained vacant ever since. After the tower began to tilt noticeably in the late 1950s, the National Park Service beginning in 1962 launched an emergency restoration effort, which included stabilizing the tower by pumping cement into its foundations. By 1969, the leaning light tower of Rock Harbor was stabilized at

about two degrees from vertical. The Park Service is continuing its restoration efforts. The surviving structures include a conical brick and stone light tower 18 feet in diameter at the base and 50 feet tall, plus a rubble masonry two-story lightkeeper's house, 29 feet square, with a gabled roof.

Passage Island Light (1882)

In a curious appropriation, Congress agreed in March 1874 to spend $18,000 for a light on Passage Island, but would build it only after the Canadian government had erected a lighthouse on Colchester Reef near the mouth of the Detroit River. As a result of this international diplomacy, construction did not begin until 1881. The work was completed the following year, and the light was first exhibited on July 1, 1882. It marks the northeastern end of Isle Royale, guiding vessels into Thunder Bay, and it is the northernmost American lighthouse on the Great Lakes. One of the keepers at Passage Island, a man named Shaw, served from 1893 to about 1910, and one of his assistants became a hero when the Canadian passenger ship *Monarch* was wrecked nearby in December, 1906. The details of the heroic rescue appear earlier in this book. The surviving buildings at Passage Island include the original light tower, keeper's house, and fog signal building. The octagonal tower is 8 feet in diameter at the base and 44 feet tall, built into one corner of the rectangular dwelling. Both are made of coursed rubble masonry construction. The Fresnel lens still in place was built by Barbier, Benard & Turenne of Paris in 1880.

Eagle River Light (1874)

The town of Eagle River was one of the earliest and most important communities to develop on the Keweenaw Peninsula in the 1840s. It was located near

SAND HILLS LIGHT

KEWEENAW WATERWAY UPPER ENTRANCE LIGHT

the Cliff Mine, the first fabulously successful copper mine in the region. The harbor at Eagle River became the shipping point for copper and the point of entry for men and supplies destined for the Cliff Mine. Congress agreed to spend $4,000 for a lighthouse at the mouth of the river in 1850, and the light was completed in 1854, but in separate actions in 1869 and 1871, Congress agreed to spend an additional $28,000 to rebuild the lighthouse ''on a proper site.'' Between 1871 and 1874, a bit less than $15,000 was actually used, with the balance returned to the Treasury. The new lighthouse then remained in service until 1908 and now serves as a private residence.

Sand Hills Light (1919)

The shoals located off Eagle River have caused many shipwrecks over the years, but the Eagle River Light-house was located too far distant to offer adequate warning, so its operation was discontinued in 1908. Then, in 1910, the ore carrier *William C. Moreland* became stranded on these shoals. The Sand Hills light, which included a powerful fog signal, was built in 1917-19 at a cost of $100,000 largely reflecting the difficulty of building a structure on this location. All materials had to be brought in by barge. The station originally had an oil vapor lamp mounted in a Fourth Order bull's-eye lens and was visible for eleven miles. William Richard Bennetts was the keeper during the entire period that this station was manned, from 1919 until 1939. The light then used an automatic acetylene gas lamp between 1939 and 1954, when the station was taken out of service. The station consists of a large brick dwelling divided into two separate residences, with a square light tower projecting from the middle of the building. It created a lens focal plane 91 feet above the level of Lake Superior.

ONTONAGON LIGHT

Keweenaw Waterway Upper Entrance Light (1950)

This structure replaced a light built in 1874 to mark the upper entrance to the Portage Lake Ship Canal, now known as the Keweenaw Waterway. The square steel tower stands 50 feet tall, but with the added height provided by the concrete base and crib structures, the lens is 82 feet above Lake Superior.

Fourteen Mile Point Light (1894)

Built in 1894 at a cost of $20,000, this light marked the long stretch of unmarked coast between Ontonagon and the upper entrance to the Keweenaw Waterway. During the last decade the light was still in service prior to its closing in 1934, the keepers were Messrs. Durframe, McGregor, and Hamm. The brick keeper's residence was a double house, with the square brick light tower attached to the front. The station originally had a Fresnel lens rotated by a clockwork mechanism, that alternately flashed red and white. On July 30, 1984, vandals started a fire which burned the lighthouse down.

Ontonagon Light (1867, 1890)

Ontonagon's fine natural harbor made this city an important shipping point for both lumber and copper by the late 1840s. Nearby mines including the Minesota

(sic), Mass, and Victoria brought in men and supplies through Ontonagon. The Federal Government built the first lighthouse there in 1852, at a cost of $5,000 with a Detroit contractor, W.F. Chittenden, completing the work. It was the fifth lighthouse erected on Lake Superior, and originally used Winslow Lewis lamps and reflectors, but it was refitted with a Fresnel lens in 1857. The first keeper was Samuel Peck. The surviving lighthouse, however, was erected in 1866-67 at a cost of $14,000, entirely replacing the 1852 light. One major alteration occurred in 1890, when a square one-story brick kitchen was added to the east facade of the building. The station remained in service until 1964, when the Coast Guard removed the lens and presented it to the Ontonagon County Historical Society. Arnold Huukie was the last keeper to run this lighthouse. The surviving building is a rectangular yellow brick residence, with a gabled roof, and an attached square brick light tower, located on the west facade of the dwelling, which stands 34 feet tall.

La Pointe Light (1891, 1897)

The first La Pointe Lighthouse was constructed in 1856, but the present structure was built in 1896-1897 at the same time the Chequamegon Point Light was being reconstructed. Operation of the old La Pointe Lighthouse was then discontinued in October, 1897, and the

new light was placed into service, with the old lens being moved to Chequamegon Point. Originally this station exhibited a Fourth Order Fresnel lens, but it is now equipped with a modern DCB-10 airport-type beacon. This structure is a pyramidal skeletal iron tower supporting a round cast iron watchroom and an octagonal cast iron lantern. The four corner posts, which are 19 feet, 5 inches from center to center, consist of sections of cast iron pipes, braced with iron pipes as horizontal struts, and diagonal tension rods of 1-3/8 inch steel. The watchroom, 9 feet, 6 inches in diameter, is surmounted by the octagonal lantern. Overall, the lighthouse is 65 feet, 3 inches tall from the base to the top of the ventilator ball, while the focal plane of the lens is 70 feet above the mean low water level of the lake. In addition, there is a rectangular frame building, 22 feet by 40 feet, with a gabled roof and corrugated iron siding that was built in 1891 to house the fog signal equipment. This station was manned until 1964.

Chequamegon Point Light (1897)

There has been a light at the west end of Long Island in Chequamegon Bay since 1858, but by the mid-1890s, this and a second light on the island were in such poor condition, that Congress appropriated $10,000 in March, 1895 to move and rebuild both stations. Construction was stopped in 1896 when the funds were exhausted and the work was not completed until an additional $1,500 was appropriated in 1897. The Fourth Order Fresnel lens from the original La Pointe lighthouse was installed in the Chequamegon Point Lighthouse in 1897, but this lens was moved again at some later date because the lighthouse now exhibits a Fifth Order Fresnel lens made by Sautter and Company of Paris. The lighthouse consists of a pyramidal skeletal tower, supported by four corner posts of 6 by 6 inch angle iron, 15 feet, 6 inches apart at the base, each resting on a concrete pier 5 inches square by 5 feet deep.

MICHIGAN ISLAND LIGHT

OUTER ISLAND LIGHT

The tower has horizontal T-shaped struts and round steel tension rods diagonal to the corner posts. The tower supports a wooden watchroom sheathed in corrugated sheet metal, measuring 10 feet by 6 inches square at the base of the watchroom. The structure also has a central weight shaft which encased the weights that originally drove the clockwork mechanism to turn the lens. Overall, the tower is 35 feet tall from the base to the top of the ventilator ball, while the lens has a focal plane 39 feet above Lake Superior.

Ashland Breakwater Light (1915)

Congress appropriated $25,000 in October, 1913 for a lighthouse to guide ships into Ashland Harbor, an important iron ore shipping center. Construction began in 1914 and the light was put into service on October 15, 1915. It originally had a Fourth Order Fresnel lens built in 1890, but the lens has since been removed. This is a rare example of the use of reinforced concrete for lighthouse construction, the only significant surviving example on the Great Lakes. Resting on a concrete pier, the structure consists of a rectangular segment at the base, measuring 20 feet by 37 feet and 14 feet high supporting a pyramidal tower 17 feet, 2 inches square at the bottom, 14 feet square at the top, and 28 feet tall, with walls 10 inches thick. It is surmounted by a round watchroom 13 feet in diameter, 8 feet high, consisting of

1/8 inch steel plates, supporting a round cast iron lantern with helical bars across the lantern glass. The focal plane of the lens is 60 feet above the mean low water level, while the lighthouse structure is 58 feet in height, measured from the base of the tower to the top of the ventilator ball.

Michigan Island Light (1857, 1880, 1930)

The light at Michigan Island was the second aid to navigation built on the Apostle Islands, an archipelago named after the Twelve Apostles, although there are more than twenty islands in the chain. The first lighthouse was at La Pointe, on Madeline Island. The light at Michigan Island helped mark the northern edge of the channel used by vessels headed west to Bayfield, Wisconsin, from ports east of the Apostles. When Ashland became a major shipping point for iron ore from the Gogebic range in the 1880s, this lighthouse marked the eastern edge of the island chain for vessels headed into Ashland. The light was erected in 1857, but the Lighthouse Board spent $6,000 in 1869, ''for renovating and relighting the lighthouse on Michigan Island,'' implying that the station had been abandoned. The original conical stone light tower has survived, as well as the lantern, which still sits atop the tower. It remained in service until 1929, when a skeletal steel tower dating from 1880 and resembling the one at Rawley Point, Wisconsin was rebuilt about 100 feet northwest of the old tower. The skeletal steel structure is 102 feet tall overall and creates a lens focal plane 170 feet above the lake. The complex also includes a two-story brick keeper's dwelling built in 1930.

Outer Island Light (1874)

As lake traffic serving the Duluth-Superior harbor increased in the early 1870s, ship owners exerted

pressure on Congress to build a light on Outer Island to mark the northern boundary of the Apostle Islands. The Lighthouse Board completed the project in 1873-74 at a cost of $40,000, erecting a major lake coast light closely resembling the one at Seul Choix Point on Lake Michigan. The station includes a large brick keeper's dwelling connected to a conical brick tower through a covered passageway. The 80 foot tower creates a lens focal plane 130 feet above Lake Superior, and now exhibits a plastic lens. The station also has a fog signal building as well as miscellaneous outbuildings. With its recent conversion to solar power, this light is also an example of the modernization process being carried out at many lights by the Coast Guard.

Devil's Island Light (1891, 1901)

This was the last major light established on the Apostle Islands and it remains one of the most powerful on Lake Superior, with a nominal range of 23 miles. Two large keeper's houses, one made of brick and the other of frame construction, that were both erected in 1891, are still standing. The station originally had a wooden light tower, which was demolished after the present tower was erected in 1901. It is a skeletal steel structure with a cylindrical steel stair tower supporting a round lantern. The tower is 71 feet tall and creates a lens focal plane 100 feet above Lake Superior.

Raspberry Island Light (1862)

In 1859, Congress appropriated a sum of $6,000 for the construction of a light at Raspberry Island and the project was completed in 1862. This light marks the West Channel through the Apostle Islands to Bayfield and points south. The large keeper's dwelling is a two-story wood-framed building with an attached wooden light tower, creating a lens focal plane 80 feet above the

lake. The station was automated and vacated in 1957, when the Coast Guard erected a new steel tower nearby to display the light.

Sand Island Light (1882)

The Sand Island Lighthouse, the westernmost light in the Apostles, was erected in 1881-82 at a cost of $18,000 to help mark the western boundary of the island chain for the growing number of vessels serving Duluth and Superior. The lighthouse is virtually identical to the ones built on Chambers Island and at Eagle Bluff, although Sand Island used sandstone as the building material, while the other two used brick. The light tower, set at a 45 degree angle into the northwest corner of the keeper's house, is square at the base, but then becomes octagonal at the second floor level. It is surmounted by a 10-sided lantern and creates a focal plane 52 feet above Lake Superior. In 1933, the Coast Guard built a skeletal steel tower for an automated light and took the old lighthouse out of service at that time.

Superior South Breakwater (Wisconsin Point) Light (1913)

Before the completion of a canal cutting across Minnesota Point, a narrow neck of land separating the mouth of the St. Louis River and Superior Bay from Lake Superior, vessels serving Duluth and Superior entered Superior Bay through a natural, but narrow and shallow channel that lies between Minnesota Point and Wisconsin Point. The first Federal lighthouse in the area was erected at Minnesota Point (Superior Entry) in 1856 at a cost of $13,700. This structure, which now lies in ruins, was rendered obsolete when the present combination lighthouse and fog signal building was erected at Wisconsin Point in 1913. The fog signal building is rectangular, with two rounded ends, and measures 25 feet by 44 feet overall. The light tower projects from the northeast corner of the fog signal building and is a concrete cylindrical structure, 42 feet tall, supporting a round cast iron lantern with helical bars. The original lens, a Fourth Order Fresnel made by Sautter, Lemonnier, and Company in 1890, is no longer present, having been replaced by a modern electric beacon. Similarly, none of the original fog signal apparatus has survived.

Duluth South Breakwater Outer Light (1901)

Despite the opposition of the state of Wisconsin and the United States War Department, the citizens of Duluth began to cut a channel between St. Louis Bay and Lake Superior in the fall of 1870, finally completing the work in 1872 after much litigation. Congress then appropriated $10,000 in March, 1870 for a light to be installed in the area, specifically "at the terminus of the

AERIAL VIEW OF TWO HARBORS LIGHT

DULUTH NORTH BREAKWATER LIGHT

Northern Pacific Railroad.'' The station, completed in 1873, included a tower on the south breakwater, but the keeper's residence was on shore. With the addition of an Inner Light on the breakwater in 1889, the Outer Light has served as a range light. The present structure was erected in 1901, after the pier was extended, and consists of a rectangular brick fog signal building, covered by a corrugated iron roof, with light tower rising from the northwest corner of the building. The round cast iron lantern is 8 feet, 4 inches in diameter with helical bars on the lantern panels. Overall, the tower stands 35 feet high and produces a focal plane 44 feet above the level of Lake Superior. The Fourth Order Fresnel lens, first exhibited on September 1, 1901, was built by Barbier and Fenestre of Paris in 1877 and was probably the lens also used at the previous lighthouse.

Duluth South Breakwater Inner Light (1901)

The first Inner Light at the Duluth South Breakwater was built in 1889 to serve as a range light in combination with the Outer Light. The present structure was begun in June, 1900, and it was completed in August 1901. The light consists of a pyramidal steel skeletal frame supporting a round watchroom and an octagonal cast iron lantern. The skeletal tower is 19 feet, 5 inches square at the base and 9 feet square at the top, where it supports the watch room, which is 9 feet in diameter and constructed of cast iron plates. The watchroom is reached from the ground level through a round stair cylinder, 6 feet in diameter, consisting of 5/16 inch cast iron plates. The stair cylinder supports the stairs, but not the watchroom and lantern. Overall, the structure is 67 feet high from the base to the top of the ventilator ball, while the focal plane of the lens is 68 feet above the mean low water level of the lake. The lantern houses a Fourth Order Fresnel lens made by Barbier and Benard of Paris in 1896 with six flash panels, each having a bull's-eye in the central drum. The original clockwork

TWO HARBORS LIGHT

mechanism which turned the lens to produce the flash is no longer in place.

Duluth North Breakwater Light (1910)

In their *Annual Report for 1908,* the Lighthouse Board described the approach to the Duluth harbor as "one of the worst and most dangerous on the whole chain of (Great) lakes" because only the South Pier of the entrance was lit and the channel leading into the harbor entrance was only 300 feet wide. To remedy this problem, they began construction of a new light at the end of the North Pier in 1909. Built at a cost of $4,000, the new light went into service in April, 1910. The conical tower, resting on a concrete base, has a steel frame enclosed by 5/16 inch riveted steel plates and measures 10 feet, 6 inches in diameter at the base, tapering to 8 feet in diameter at the top. The octagonal lantern has an inscribed diameter of 7 feet, 6 inches and

is constructed of cast iron. Overall, the tower is 37 feet high from the base to the top of the ventilator ball, but the focal plane of the lens is 43 feet above Lake Superior. The lens is a Fifth Order Fresnel consisting of four panels of 90 degrees, with a central drum of 5 elements, with the segments above and below the central drum containing 6 prisms and 3 prisms respectively. The lens was made by Henri Le Paute of Paris in 1882.

Two Harbors Light (1892)

At the time the Two Harbors Light Station was completed in 1892, this was the major shipping point for the iron ore taken from the Mesabi Range, far surpassing Duluth in importance. The keeper's house is a two-story red brick building 35 feet square, with gabled roofs. The attached light tower, also made of brick, is 12 feet, 9 inches square, and stands 49 feet tall from the base to the top of the ventilator ball, with the

SPLIT ROCK LIGHT

FOG SIGNAL BUILDING AND TOWER AT SPLIT ROCK

lens focal plane 78 feet above Lake Superior. The tower supports an octagonal cast iron lantern. The original Fourth Order Fresnel lens was replaced in 1970 with a pair of rotating airport-type electric beacons. The site also contains the assistant keeper's house, a frame L-shaped building with gabled roofs. Also standing are the original fog signal house, a rectangular building with a gabled roof, measuring 22 by 40 feet, as well as the old oil house which is also made of brick.

Split Rock Light (1910)

The Lighthouse Board decided to build a light at this location for two reasons. First, compass readings were often unreliable because of the magnetic interference created by large iron ore deposits in the vicinity, to say nothing of the ore that was being carried by vessels in the area. As a result, several shipwrecks occurred on these rocky coasts, particularly in foggy weather. The second reason was the extreme depth of the lake close to shore. This made it difficult for the mariners to sound for depth at a safe distance from the rocky and hazardous shoreline. The light station was virtually completed during the summer of 1909 at a cost of $72,541, but it did not begin operating until 1910. The construction of Split Rock Light was a triumph of engineering and logistics because all the materials and men had to be brought to the site over Lake Superior and then hoisted

up the face of a cliff some 124 feet in height. The octagonal brick light tower stands 54 feet tall overall, but created a lens focal plane 168 feet above the lake. It was equipped with an incandescent oil vapor lamp rated at 370,000 candlepower, mounted in a Third Order lens. The official range was 22 miles. In 1939, the station was wired for electricity, and for the next 30 years, a 1,000 watt bulb supplied the light source. The station includes a fog signal building, three barns, an oil storage building, and three two-story brick dwellings to accommodate the three keepers' families that initially manned the station. The Coast Guard decommissioned the lighthouse in 1969. The Minnesota Historical Society acquired the property in 1975, and today operates the complex as a historic site.

Grand Marais Light (1922)

The first lighthouse at Grand Marais was erected in 1885, but the present structure was built in 1922. It consists of a pyramidal steel frame, 25 feet high, 13 feet square at the base and 9 feet square at the top, the upper portion encased in steel plates to form the watchroom. It supports an octagonal cast iron lantern, which contains the lens. Overall, the structure is 34 feet tall from the base to the top of the ventilator ball and results in a lens focal plane 38 feet above the mean low water level of Lake Superior. The lens is a Fifth Order Fresnel, bearing the markings, ''Sautter & Co., Constructeurs, Paris,'' but it is not marked with a date. However, it is probably the original lens, installed in 1885. □

EPILOGUE

Much of the importance of Great Lakes maritime history, including the role of lighthouses and their keepers, seems distant and perhaps is entirely lost on people living in the late twentieth century. We are part of an era in which innovations such as the Interstate Highway System, supersonic air travel, the Space Shuttle, atomic power, electronic computers, laser technology, and other marvels are almost greeted with indifference by a technically sophisticated public which expects major new conquests of nature to occur routinely. Even the people living on or near the Great Lakes easily forget the significance or even the existence of these great inland seas. The Great Lakes offer more variety of dangers to mariners and their vessels than any ocean. Their violent storms, including winter gales, as well as fog and ice, match anything found on the seven seas. The Lakes' many narrow and shallow passages, combined with the great volume of traffic, have produced another set of dangers. The thousands of men and ships that the Lakes have claimed over the past two centuries are a grim monument to the hazards of traveling on them. Modern technology has reduced, but certainly not eliminated the risks. The fate of many ships, such as the *Edmund Fitzgerald*, which went down without even sending a distress signal, is evidence of the continuing dangers. We should also not lose sight of the enormous influence that Lake commerce has exerted on the history of this region. The development of the agricultural, forest, and mineral resources of the states bordering on the Lakes would have been much different without the existence of cheap and reliable water transport from the 1840s on. The subsequent industrial development of the region was similarly affected.

With few exceptions, the U.S. Lighthouse Service adequately served the interests of the owners, captains, and crews of vessels on the Great Lakes, as well as the interests of the shippers and the general public. The building, maintenance, and staffing of the lighthouses, light vessels, fog signals, buoys, and other aids to navigation were a Federal responsibility from the beginning, and this helped produce a reasonably consistent and rational system. The Lighthouse Service seriously considered the complaints and suggestions of shipowners, including the views of the Lake Carriers' Association, as did Congress, so the system was responsive to the legitimate needs of navigation. Lighthouses were not only constructed as needed, but in general, were competently designed, built, and maintained. Except for the choice of lamps and lenses during Stephen Pleasonton's administration, the Lighthouse Service has been right on the technological frontier most of the time. While it is always possible to cite individual, isolated examples of gross incompetence or dishonesty in any government agency, there is no evidence that these were ever serious problems in the Lighthouse Service. After the Lighthouse Board took control of the system in 1852, helped by the later development of Civil Service requirement, the quality of personnel as well as facilities improved noticeably. This tradition of excellence continued under the Bureau of Lighthouses beginning in 1910 and then, from 1939 onwards, under the United States Coast Guard.

The history of Great Lakes lighthouses over the past five or six decades has been a history of automation and abandonment. The trend began with the invention of automatic lights using acetylene gas, but automation did not take giant strides until the spread of electric service in the 1920s and after. With the development of reliable portable generators and improved storage batteries, even the most isolated light station could be automated. The traditional light tower of wood, stone, or brick quickly became an artifact of the past. The skeletal steel tower, sometimes encased in steel plates, has been practically the only design used in new construction since the 1920s. Finally, the development of radar and radiobeacons has given vessels more accurate and far-reaching warnings than any lighthouse could and has made many lights obsolete. As a result of these developments, the Bureau of Lighthouses and the Coast Guard have automated hundreds and entirely abandoned scores of light stations on the Great Lakes. Still, navigation is better served today than ever before.

Ironically, the disappearance of manned light stations and the numerous abandonments have made it easier for people living in the region, as well as tourists, to visit lighthouses and understand their history. In the last two or three decades, the Federal government or the Coast Guard has sold or given away many lighthouses and their equipment to the state and to several local governments, as well as to a few private historical societies. Museums of maritime history have benefited, including the Great Lakes Historical Society Museum in Vermilion, Ohio, and the Dossin Museum on Belle Isle in Detroit. Numerous museums have appeared in or near abandoned lighthouses, including ones at Pointe Aux Barques, Sturgeon Point, Presque Isle, Mackinac Point, South Manitou Island, and at White River (Whitehall), all in Michigan's Lower Peninsula. Important lighthouse museums now exist in Michigan City, Indiana, at the Grosse Point Lighthouse in Evanston, Illinois, and at the Eagle Bluff Light in Door County, Wisconsin. Finally, on Lake Superior, one can visit museums at Whitefish Point, Copper Harbor, Eagle Harbor, and at the Split Rock Lighthouse in Minnesota. Another sign of the growing interest in this part of our history was the founding, in 1982, of the Great Lakes Lighthouse Keepers Association, an organization dedicated to preserving lighthouse history and lore. The Association has separate chapters for each of the Lakes and publishes a quarterly, *The Beacon*.

The lightkeepers of the nineteenth and early twentieth centuries were exceptional characters in the drama of American industrialization, urban growth, and the creeping depersonalization of work and life. Their labor was solitary and isolated from the main currents of American life. Keepers were highly individualistic, independent, and self-reliant men and women, part of a vanished American frontier and culture. Their lives were difficult in large part because of the isolation, monotony, and boredom they had to endure. They were unsung heroes, and in a narrow sense, as lightkeepers, they were their brothers' keepers most of the time. They sometimes faced serious dangers from storms and ice, and on rare occasions, became heroes and heroines to rescue sailors in distress. They are a dying breed nationally, having already succumbed on the Great Lakes. We can learn much about the human spirit from them. The poem by Edgar Guest on the following page perhaps best sums up the bittersweet feelings many of us have in witnessing the passing of this era:

THE LIGHTKEEPER WONDERS

The light I've tended for 40 years
is now to be run by a set of gears.
The Keeper said, And it isn't nice
To be put ashore by a mere device.
Now, fair or foul the winds that blow
Or smooth or rough the sea below,
It is all the same. The ships at
night will run to an automatic light.

That clock and gear which truly turn
Are timed and set so the light shall burn.
But did ever an automatic thing
set plants about in early Spring?
And did ever a bit of wire and gear
A cry for help in the darkness hear?
Or welcome callers and show them through
The lighthouse rooms as I used to do?

'Tis not in malice these things I say
All men must bow to the newer way.
But it's strange for a lighthouse man like me
After forty years on shore to be.
And I wonder now—will the grass stay green?
Will the brass stay bright and the windows clean?
And will ever that automatic thing
Plant marigolds in early Spring?

Edgar Guest

SUNSET OVER LUDINGTON LIGHT, LAKE MICHIGAN

PHOTOGRAPHERS' STATEMENT

What would compel a family to undertake a three-year odyssey to search out and photograph more than 200 Great Lakes lighthouses? In part, it is our shared love of the Great Lakes themselves—the vast freshwater seas that equal the oceans in beauty and danger. Certainly, the lighthouses of these inland seas are significant as more than just historical monuments, as more than expressions of grace and beauty in picturesque settings. Most importantly, they stand as reminders of a past marked by heroism and devotion to duty. These empty buildings would not capture our imagination and fascination if not for the lives of the men and women who worked and lived in them.

Our journey to the sites of these human dramas has been a delightful adventure. With our daughters, Jennifer and Kristen, we have explored the shores and islands of the Great Lakes. In the Beaver Island archipelago, we visited small islands with clouds of soaring gulls, and clearings filled with colorful wildflowers (but we also endured biting flies and flourishing poison ivy!). In the Straits of Mackinac, we reached St. Helena Island by a roller-coaster ride over swells that dwarfed us in our small boat. At Crisp's Point, we became lost in a maze of sandy trails, hiking miles out of our way—along remote Lake Superior shores with scenery that made it worth the extra effort. And, having promised the girls sherbet after a long day's drive, we stopped in Paradise, Michigan—only to be informed by a local innkeeper that "there is no sherbet in Paradise!"

We often watched the silver-white streaks of the northern lights shimmering in the night skies, and slept in our camper or tent, lulled by the sounds of the waves on the shore and the mournful foghorns.

On remote Passage Island (near Isle Royale), we learned of life in a lighthouse from Anna Hoge, the lightkeeper's daughter who lived there as a child over 40 years ago. We fancied ourselves "lightkeepers for a day" when we were allowed to stay overnight in the lighthouse.

At Rock of Ages, we also briefly experienced life in a lighthouse, and were spellbound by the 10-foot-high second order Fresnel lens. That evening, watching as the lantern room was filled with a light show of rainbows, and listening to the great lens rotating in its bed of mercury, we felt a sense of timelessness—as all around us were the very sights and sounds experienced long ago by the keepers of the light.

We are grateful to the many people we met in our travels who have given us direction and assistance in a variety of ways; we especially want to thank the Coast Guard units that helped us locate certain lighthouses, and assisted us in reaching some of the more remote ones. We are also grateful to the lightkeepers and their families who have shared memories with us, giving added insights into the lightkeepers' lives.

In returning to many lighthouses, we have been impressed with how quickly things change. Fresnel lenses have been replaced, outbuildings destroyed, lighthouses have been vandalized, and some have even burned. Thus, it became increasingly important to us to create a photographic collection of the U.S. Great Lakes lighthouses that is as complete as possible—to document and preserve this significant portion of our Great Lakes heritage before more changes occur. (A note about our photographs: We do not use special effects filters or "trick" photography; rather, we prefer to capture

PHOTOGRAPHERS AND FAMILY AT PASSAGE ISLAND LIGHT

the unique and dramatic colors and lighting that are created by nature.)

Through our travel experiences, many hours of research, and contacts with lightkeepers and their families, we have developed not only a sense of the wonderful beauty that the Great Lakes hold, but also a deep appreciation of the rich history of the Great Lakes, which includes the lighthouses. Although our work for this book is completed, our Great Lakes odyssey continues on.

Many people share our fascination with lighthouses and respect for the keepers of the lights. Lighthouses are often used to symbolize many things; but most importantly, these monoliths are tangible symbols of man's caring for his fellow man. They remind us of the great drama that goes on when man pits himself against nature's forces. It is the hopeful, positive story of people isolated in remote places—for the sole purpose of trying to aid their fellow man and warn him of danger. And when there were shipwrecks nearby, they went down to the water, or out in small boats to bring in the survivors—at great risk to their own lives.

As the last lights on the Great Lakes were automated and unmanned, we felt privileged and yet sad to have witnessed the end of an era marked by bravery and devotion to duty. Some lighthouses have simply been abandoned—to be eventually reclaimed by the natural elements that initially justified their existence. Some are being reclaimed and restored by individuals and groups committed to preserving their history. But many of the lighthouses still stand—reminders of their rich past—continuing their lonely vigil, and giving guidance and warning to the sailors of these great inland seas.

Ann and John Mahan
Stevensville, Michigan

199

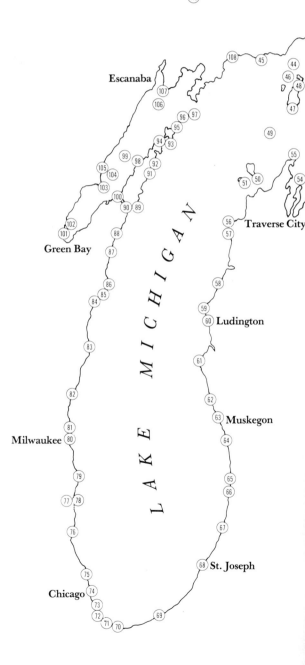

1. West Sister Island Light: At southwest end of West Sister Island, entrance to Maumee Bay, east of Toledo, Ohio.

2. Manhattan Range Lights: North bank of Maumee River, east of Summit Street, Toledo, Ohio.

3. Toledo Harbor Light: North side of outer entrance to Maumee Bay and Toledo Harbor, Toledo, Ohio.

4. Turtle Island Light: Ruins on Turtle Island, in Maumee Bay.

5. Detroit River Light: At entrance to Detroit River, Lake Erie.

6. Grosse Ile North Channel Front Range Light: At Lighthouse Point, east side of Grosse Ile, on the Detroit River.

7. Detroit Lighthouse Depot: Foot of Mount Elliot Street, Detroit.

8. William Livingstone Memorial Light: East end of Belle Isle, Detroit, on the Detroit River.

9. Windmill Point Light: Southwest corner of U.S. Marine Hospital Reservation, Detroit, on Lake St. Clair.

10. Lake St. Clair Light: West side of shipping channel, at the turn, Lake St. Clair.

11. St. Clair Flats Old Channel Range Lights: At the head of the old channel, northeast end of Lake St. Clair.

12. Peche Island Rear Range Light: At St. Clair River, Marine City.

13. Lightship Huron: Pine Grove Park at St. Clair River, Port Huron.

14. Fort Gratiot Light: Omar and Garfield Streets, Port Huron.

15. Port Sanilac Light: East side of Lake Street, Port Sanilac.

16. Harbor Beach (Sand Beach) Light: North side of breakwater entrance, Harbor Beach.

17. Pointe Aux Barques Light: Lighthouse Road, Huron Township.

18. Port Austin Reef Light: East end of Port Austin Reef.

19. Charity Island Light: On Big Charity Island, middle of entrance to Saginaw Bay.

20. Gravelly Shoal Light: On outer end of shoal, three miles southeast of Point Lookout.

21. Saginaw River Rear Range Light: Coast Guard Street, west side of Saginaw River, Bay City.

22. Tawas Point Light: On Tawas Point, at the end of Tawas Point Road, East Tawas.

23. Sturgeon Point Light: End of Point Road, off Lakeshore Drive, north of Harrisville.

24. Alpena Light: Northeast side of entrance to Thunder Bay River, at Thunder Bay.

25. Thunder Bay Island Light: Southeast shore of Thunder Bay Island.

26. Middle Island Light: East side of Middle Island.

27. Old Presque Isle Light: East side of Presque Isle Harbor. **Presque Isle Light:** North end of Presque Isle. **Presque Isle Range Lights:** West shore of Presque Isle Harbor.

28. Forty Mile Point Light: Presque Isle County Park, north of Rogers City, Hammond Bay.

29. Cheboygan River Range Front Light: First Street at Water Street, Cheboygan. **Cheboygan Crib Light:** Entrance to Black River. **Cheboygan Light:** Ruins at Cheboygan Point.

30. Fourteen Foot Shoal Light: Northwest of Cheboygan Point.

31. Poe Reef Light: Four miles northeast of Cheboygan.

32. Bois Blanc Island Light: On Lighthouse Point, northeast side of Bois Blanc Island, Lake Huron.

33. Spectacle Reef Light: Ten miles east of Bois Blanc Island, Lake Huron.

34. Round Island Light: Northwest shore of Round Island, south of Mackinac Island, Lake Huron.

35. Martin Reef Light: Martin Reef, four miles south of Cadogan Point, Lake Huron.

36. DeTour Point Light: End of DeTour Reef, west side of entrance to St. Mary's River, Lake Huron.

37. Old Mackinac Point Light: South side of the Straits of Mackinac, in Michilimackinac State Park.

38. McGulpin's Point Light: South side of the Straits of Mackinac, three miles west of Mackinaw City.

39. Waugoshance Light: Northwest of Waugoshance Island, west of the Straits of Mackinac, Lake Michigan.

40. Skillagalee (Ile Aux Galets) Light: On Skillagalee Rock, southwest of Waugoshance Island, Lake Michigan.

LIGHTHOUSE MAP AND LOCATIONS

41. White Shoal Light: Northwest of Waugoshance Island, twenty miles west of Mackinac Point.

42. Gray's Reef Light: West of Waugoshance Island, Lake Michigan.

43. St. Helena Island Light: Southeast end of St. Helena Island, northwest of the Straits of Mackinac.

44. Lansing Shoal Light: Forty miles west of Mackinaw City, Lake Michigan.

45. Seul Choix Point Light: At Seul Choix Point, eighteen miles east of Manistique, Lake Michigan.

46. Squaw Island Light: On Squaw Island, six miles north of Beaver Island.

47. Beaver Island (Beaver Head) Light: South shore of Beaver Island, between Nicksau's Point and Appleby's Point.

48. Beaver Island Harbor (St. James) Light: North side of entrance to Beaver Island Harbor.

49. South Fox Island Light: Southeast end of South Fox Island.

50. North Manitou Shoal Light: South end of shoal, off North Manitou Island.

51. South Manitou Island Light: Southeast point of South Manitou Island.

52. Little Traverse (Harbor Point) Light: At Harbor Point, north side of Little Traverse Bay, in Harbor Springs.

53. Charlevoix South Pier Light: Outer end of south pier.

54. Mission Point (Old Mission Point) Light: At the tip of Old Mission Peninsula, Grand Traverse Bay.

55. Grand Traverse (Cat's Head Point) Light: North tip of Leelanau Peninsula.

56. Point Betsie Light: On Point Betsie, north of Frankfort.

57. Frankfort North Breakwater Light: Outer end of north breakwater.

58. Manistee North Pierhead Light: Outer end of north pier.

59. Big Sable Point (Grand Point au Sable) Light: On Big Sable Point, eight miles north of Ludington.

60. Ludington North Pierhead Light: Outer end of north pier, at harbor entrance.

61. Little Sable Point (Petite Pointe au Sable) Light: On Little Sable Point, south of Silver Lake.

62. White River Light: At entrance to White Lake, in Whitehall.

63. Muskegon South Pier Light: Outer end of south pier. **Muskegon South Breakwater Light:** South end of breakwater.

64. Grand Haven South Pier Inner Light: Approximately 600 feet east of pierhead. **Grand Haven South Pierhead:** End of south pier.

65. Holland Harbor (Black Lake) Light: South Pier, at entrance to Lake Macatawa.

66. Saugatuck (Kalamazoo River) Light: Mouth of Kalamazoo River.

67. South Haven South Pier Light: Outer end of south pier.

68. St. Joseph North Pier Outer Light: Outer end of north pier. **St. Joseph North Pier Inner Light:** Middle of north pier.

69. Michigan City Light: At the bend in the harbor, in Washington Park. **Michigan City East Pier Light:** Outer end of east pier. **Michigan City Breakwater Light:** North end of breakwater.

70. Gary Breakwater Light: End of breakwater.

71. Buffington Breakwater Light: East end of north breakwater.

72. Indiana Harbor East Breakwater Light: North end of east breakwater.

73. Calumet Harbor (South Chicago) Light: Southeast end of north breakwater.

74. Chicago Harbor Southeast Guidewall Light: Outer end of wall at entrance to the lock, Chicago River entrance into the harbor. **Chicago Harbor Light:** South end of north breakwater.

75. Grosse Point Light: 2535 Sheridan Road, Evanston, Illinois.

76. Waukegan Harbor (Little Fort) Light: Outer end of south pier.

77. Kenosha (Southport) Light: On Simmons Island, north side of harbor entrance.

78. Kenosha Pierhead Light: End of north pier.

79. Wind Point (Racine Point) Light: Windridge Drive, north of Racine.

80. Milwaukee Pierhead Light: Outer end of north pier. **Milwaukee Breakwater Light:** Southeast end of north breakwater.

81. North Point Light: Wahl Street at Terrace, in Milwaukee.

82. Port Washington Breakwater Light: Southeast end of north breakwater.

83. Sheboygan Breakwater Light: Southeast end of breakwater.

84. Manitowoc Breakwater Light: Southeast end of north breakwater.

85. Two Rivers Light: At mouth of the river in Two Rivers.

86. Rawley Point (Twin River Point) Light: North of Two Rivers, in Point Beach State Forest.

87. Kewaunee Pierhead Light: Outer end of south pier.

88. Algoma Pierhead Front Light: End of north pier.

89. Sturgeon Bay Canal North Pierhead Light: End of north pier, east entrance to Sturgeon Bay Canal.

90. Sturgeon Bay Canal Light: North side of east entrance to Sturgeon Bay Canal.

91. Bailey's Harbor Range Lights: At harbor entrance, on Door Peninsula.

92. Cana Island Light: Northeast of Bailey's Harbor, on east side of Cana Island.

93. Pilot Island Light: On Pilot Island, east entrance to Porte Des Morts passage.

94. Plum Island Rear Range Light: Southwest side of Plum Island.

95. Potawatomi (Rock Island) Light: Northwest point of Rock Island.

96. St. Martin Island Light: Northeast corner of St. Martin Island.

97. Poverty Island Light: South side of Poverty Island.

98. Eagle Bluff Light: On west part of the bluff, on Shore Road, in Peninsula State Park.

99. Chambers Island Light: Northwest corner of Chambers Island.

100. Sherwood Point Light: South side of Sturgeon Bay Canal entrance, on Green Bay.

101. Green Bay Entrance Light: West side of channel.

102. Long Tail Point Light: At Long Tail Point, north of Green Bay.

103. Peshtigo Reef Light: Southeast point of shoal.

104. Green Island Light: Ruins on Green Island, five miles southeast of Menominee.

105. Menominee North Pier Light: End of north pier.

106. Minneapolis Shoal Light: Ten miles southeast of Escanaba.

107. Peninsula Point Light: On Peninsula Point.

108. Manistique East Breakwater Light: End of east breakwater.

109. Round Island Light: On Round Island, northwest of Lime Island, lower end of St. Mary's River.

110. Cedar Point (Round Island Point) Rear Range Light: On St. Mary's River, four miles east of Brimley.

111. Point Iroquois Light: On Iroquois Point, six miles northwest of Brimley.

112. Whitefish Point Light: On Whitefish Point, north of Paradise.

113. Crisp's Point Light: Thirteen miles west of Whitefish Point.

114. Grand Marais Harbor Range Lights: At outer end of pier, and 2,610 feet from outer end of pier.

115. Au Sable Point Light: On Au Sable Point, west of Grand Marais.

116. Grand Island North Light: On the northwest point of Grand Island.

117. Grand Island East Channel Light: Southeast tip of Grand Island.

118. Grand Island West Channel Light: On M-28, east of Christmas.

119. Munising Range Lights: At the north and south ends of Hemlock Street, in Munising.

120. Marquette Breakwater Outer Light: Outer end of breakwater. **Marquette Harbor Light:** On north point of Marquette Harbor.

121. Presque Isle Harbor Breakwater Light: End of breakwater.

122. Granite Island Light: Northwest tip of Granite Island.

123. Stannard Rock Light: North end of reef.

124. Big Bay Point Light: At Big Bay Point, twenty-four miles northwest of Marquette.

125. Huron Island Light: Northeast side of West Huron Island, three miles offshore.

126. Sand Point Light: On Sand Point, northwest of L'Anse and east of Baraga.

127. Portage Lake Lower Entrance Light: On east side of the channel, at end of the breakwater.

128. Portage River (Jacobsville) Light: One mile east of Portage Entry.

129. Mendota (Bete Grise) Light: South side of entrance to Mendota Canal, on Mendota Point.

130. Gull Rock Light: On Gull Rock, west of Manitou Island.

131. Manitou Island Light: East point of Manitou Island.

132. Copper Harbor Light: East point of harbor entrance. **Copper Harbor Range Lights:** South side of harbor.

133. Eagle Harbor Light: West end of Eagle Harbor. **Eagle Harbor Range Light:** South end of Eagle Harbor.

134. Rock of Ages Light: Off west end of Isle Royale.

135. Isle Royale (Menagerie Island) Light: Northeast end of Menagerie Island, south of Isle Royale.

136. Rock Harbor Light: At Middle Island Passage, south side of Isle Royale.

137. Passage Island Light: Southwest point of Passage Island.

138. Eagle River Light: South bank of Eagle River.

139. Sand Hills Light: At Five Mile Point.

140. Keweenaw Waterway Upper Entrance Light: Outer end of east breakwater.

141. Fourteen Mile Point Light: Fourteen miles northeast of Ontonagon.

142. Ontonagon Light: South bank of Ontonagon River.

143. La Pointe Light: North shore of Long Island, in Chequamegon Bay.

144. Chequamegon Point Light: West end of Long Island, in Chequamegon Bay.

145. Ashland Breakwater Light: West end of breakwater, in Chequamegon Bay.

146. Michigan Island Light: South end of Michigan Island.

147. Outer Island Light: North point of Outer Island.

148. Devil's Island Light: North end of Devil's Island.

149. Raspberry Island Light: Southwest point of Raspberry Island.

150. Sand Island Light: On north point of Sand Island.

151. Superior South Breakwater (Wisconsin Point) Light: Outer end of south breakwater.

152. Duluth South Breakwater Outer Light: Outer end of south breakwater. **Duluth South Breakwater Inner Light:** Inner end of south breakwater. **Duluth North Breakwater Light:** Outer end of north breakwater.

153. Two Harbors Light: On the point between Agate and Burlington Bays.

154. Split Rock Light: On Split Rock, thirty-eight miles northeast of Duluth.

155. Grand Marais Light: Outer end of east breakwater.

FURTHER READING

Adams, E.P. "The Lighthouse System of the United States." *Journal of Engineering Societies*, XII (October,1893), pp. 509-531.

Adams, William Henry Davenport. *Lighthouses and Lightships: A Description and Historical Account of Their Mode of Construction and Organization.* New York, New York, 1870.

Adamson, Hans Christian. *Keepers of the Lights.* New York, New York, 1955.

Findlay, Alexander George. *The Lighthouses of the World.* London, England, 1899.

Havighurst, Walter. *The Long Ships Passing: The Story of the Great Lakes.* New York, New York, 1975.

Heap, David Porter. *Ancient and Modern Lighthouses.* Boston, Massachusetts, 1889.

Holland, Francis. *America's Lighthouses: Their Illustrated History Since 1716.* Brattleboro, Vermont, 1972.

Johnson, Arthur Burges. *The Modern Light-House Service.* Washington, D.C., 1889.

Landon, Fred. *Lake Huron,* New York, New York, 1944.

Law, William Hainstock. *Among the Lighthouses of the Great Lakes.* Detroit, Michigan, 1908.

Mansfield, John Brandt, editor. *History of the Great Lakes.* Chicago, Illinois, 1899.

Nute, Grace Lee. *Lake Superior.* New York, New York, 1944.

O'Brien, Thomas Michael. *Guardians of the Eighth Sea: A History of the U.S. Coast Guard On the Great Lakes.* Cleveland, Ohio, 1977.

Putnam, George Rockwell. *Lighthouses and Lightships of the United States.* Boston, Massachusetts, 1917.

_____. *Sentinels of the Coast: The Diary of a Lighthouse Engineer.* New York, New York, 1937.

Quaife, Milo M. *Lake Michigan.* New York, New York, 1944.

Snow, Edward Rowe. *Famous Lighthouses of America.* New York, New York, 1955.

Stevenson, D. Alan. *The World's Lighthouses Before 1820.* London, England, 1959.

United States Coast Guard. *Guide to Historically Famous Lighthouses of the United States.* Washington, D.C., 1939.

United States Lighthouse Board. *Instructions and Directions to Light-House and Light Vessel Keepers of the United States.* Washington, D.C., Third Edition, 1858.

Index

206

207